SpringerBriefs in Biochemistry and Molecular Biology

For further volumes:
http://www.springer.com/series/10196

Anirban Banerji

Fractal Symmetry
of Protein Exterior

 Springer

Anirban Banerji
Bioinformatics Centre
University of Pune
Pune
India

ISSN 2211-9353   ISSN 2211-9361   (electronic)
ISBN 978-3-0348-0653-4   ISBN 978-3-0348-0654-1   (eBook)
DOI 10.1007/978-3-0348-0654-1
Springer Basel Heidelberg New York Dordrecht London

Library of Congress Control Number: 2013934381

Printed on acid-free paper

Springer is part of Springer Science+Business Media (www.springer.com)

This book is dedicated to my parents,
Leena Banerji and Swapan Banerji

Preface

The essential question that fractal dimensions attempt to answer is about the *scales* in Nature. This is not exactly a new question. From Euclid to Poincare to Mandelbrot, many philosophers, mathematicians, physicists had thought about transformation, dilation and contraction of scales. This small book will attempt to investigate the diverse facets of the question: *Amongst the various protein surface properties, which ones are scale-invariant*? A detailed exploration of this problem will perhaps help us to investigate utilitarian questions like: *How can we apply the acquired knowledge about the nature of protein surface roughness to drug-discovery and other practical branches of science*?

Discussions on scale-invariance lead us to fractals. Unfortunately, elaborations on what are fractals and what are their historical roots—could not be made a part of this book. Interested readers may find these topics in my other book 'Fractal Symmetry of Protein Interior', which is being published by the same publisher. Any formal investigation of protein shape and protein surface necessarily leads us to the paradigm of (complex) mathematics. Studying scale-invariance of protein exterior properties, especially, demands a good command of mathematical constructs. But it was impossible to review these constructs within the tiny volume of this book. Thus, instead of being mathematically pedagogic, a qualitative approach has been adopted throughout the course of this book with clear onus on discussions related to biological aspects of the problem.

The book attempts to collate and classify the various fractal dimension-based approaches that have been employed over the years to study protein exterior biophysical properties, into distinct clusters. It then presents an account of cases where fractal dimension-based methodologies have successfully contributed to protein exterior research. In this context, a thorough documentation of accurate predictions made from the spectrum of fractal dimension-based studies has been provided. However, fractals are no panacea, and they cannot suggest magic solutions to all the problems of protein exterior biophysics. Consequently, the third aim of this book was to examine the intrinsic limitations of fractal dimension-based measures. Finally, with a balanced assessment of the entire framework, the book attempted to identify some of the outstanding questions, where application of fractal dimension-based investigations may prove to be helpful in deciphering deep and unexpected facets of protein exterior organization.

This book wouldn't have come into being without the direct or indirect support from the countably infinite number of well-wishers of mine. I thank them all. I take this opportunity to thank all those who asked me questions about fractals and proteins over the years and especially to those whom I could not answer for my lack of knowledge and/or conception. These questions stayed with me and forced me to spend more time in the library and in my work. In a way, it is through these interactions that I learned whatever little I've learned about fractals and proteins—Thank you all.

This preface will be incomplete if I do not mention the roles of the editors, Dr. Beatrice Menz and (later on) Dr. Jutta Lindenborn during the course of writing the book. They were always supportive, always encouraging. I thank Dr. Menz especially for entrusting me with the responsibility of writing this book.

There's nothing more helpful than to listen to constructive criticisms; thus, if you feel like criticizing and/or discussing and/or (in the unlikely case) appreciating any aspect of whatever that is written in the book, please drop me an email anirbanab@gmail.com. I'll be delighted to talk to you and learn from you.

Pune, India   Anirban Banerji

Contents

Chapter 1
Introduction

Abstract Here we introduce ourselves to fractal dimensions, proteins, protein exteriors, and to methods that study roughness of any surface. Although fractal concepts, to quite an extent are mathematical, these expositions are avoided in this chapter (and in the book in general); unless the use of some expressions becomes absolutely obligatory. To help the interested readers, numerous references are suggested. Various misconceptions often plague students and researchers in the field of fractals;—a small attempt, therefore, is made to point out some of the most notable misconceptions. Though surface roughness studies are not exactly new, deciphering patterns in roughness of protein surface is a field still in infancy. Hence a short introduction to protein exterior is presented here, alongside some pertinent aspects of roughness studies.

1.1 Prelude to Introduction to Fractals

Before we start talking about origin of the word 'fractal', before we define fractal dimensions, and before we talk about the fractal organization of protein exterior, let us make fractal's acquaintance through a distinguished property of its.

One of the most essential properties of a fractal is that its structure appears to be the same when it is examined on any scale of magnification. Though these words are effortlessly spoken or written, they imply a somewhat peculiar property, called self-similarity. In order to appreciate the relevance of fractal dimension in protein exterior studies, it is absolutely necessary that we get familiar with self-similarity.

There is a reason why I claimed it to be somewhat peculiar. Just pause a while and think of it; when looking out for the finer details of a structure, a kid attempts to examine it under a magnifying glass, a grown-up does it under a microscope. We do so because the usefulness of these magnifying devices lie in the very fact that they make magnified structures look different from what are seen by the naked eye. However, while observing a fractal, all these devices will fail! Fractals embody the same level of complexity at all the level of magnification. If that does not sound sufficiently startling, surely the fact that many natural shapes are

A. Banerji, *Fractal Symmetry of Protein Exterior*, SpringerBriefs in Biochemistry and Molecular Biology, DOI: 10.1007/978-3-0348-0654-1_1, © The Author(s) 2013

fractal—seems amazing. So, even without knowing what fractals are we realize that these things are truly different from what we have learnt in our school textbooks. From the perspective of geometry, the foremost attraction of fractal dimension stems from its ability to quantify the symmetry in the extent of irregularity in natural shapes and natural features,—something that the traditional Euclidean geometry-based tools fail to perform.

As mentioned earlier, it seems truly intriguing that Nature has organized itself with very many types of fractal objects, proteins being one of them. Distributions of various protein properties (of interior and exterior) embody unmistakably self-similar symmetries. But before we get into proteins, let us take a step backward and talk a bit more about fractals.

Why is geometry often described as "cold" and "dry"? One reason lies in its inability to describe the shape of a cloud, a mountain, or a tree. Clouds are not spheres, mountains not cones, coastlines are not circles, and bark is not smooth, nor does lightning travel in a straight line.

—Said 'The Fractal Geometry of Nature', a 1982 book written by Mandelbrot (1982). These were courageous and prophetic doubts (and statements). But Mandelbrot was thinking in these lines for a quite a while. In his 1967 paper (Mandelbrot 1967), he talked about Similarity dimension. Here, Mandelbrot formalized the work of Richardson (1961), who noticed that the length of a coastline depends on the unit of measurement used. Thinking along this line, in one sharp piece of realization Mandelbrot recognized that geometrically self-similar figures will seldom be found in nature, but nature will embody an inexhaustible number of cases with statistical forms of self-similarity. Though the idea of describing natural phenomena by studying statistical scaling laws was not exactly new (Bachelier 1900; Kolmogorov 1941; Mandelbrot 1963), the ubiquity of statistical self-similarity and the reasons why fractal geometry will be more appropriate in describing the nature than classical Euclidean geometry has been,—is elaborated in detail in his 1982 book (Mandelbrot 1982). Interested readers, while searching for discussions on topological dimension, geometric self-similarity, and statistical self-similarity—may also benefit from reading (Kaye 1994). Figures 1.1 and 1.2 show the difference between a case of geometric self-similarity and statistical self-similarity. The Fig. 1.1 shows an example of exact (geometrical) self-similarity, displayed vividly by the 'Koch-curve'. Figure 1.2, in contrast, is an example of statistical self-similarity, where the roughness profile of the Earth's surface is shown. Both diagrams have been reproduced here courtesy of Professor Paul Bourke (http://paulbourke.net/fractals/fracdim/).

How to formally define a fractal?—I'm afraid; this question cannot be answered in this book; because a discussion on formal definition of fractals requires a thorough foundation on dimensional analysis among many other mathematical prerequisites. A somewhat-rigorous and possibly-formal way of answering the question 'what is a fractal?' can be: it is a fractal that represents the attractor of an iterated function system on some compact subset of a complete metric space— Unfortunately, this sounds a bit cataclysmic and certainly does not appear to be

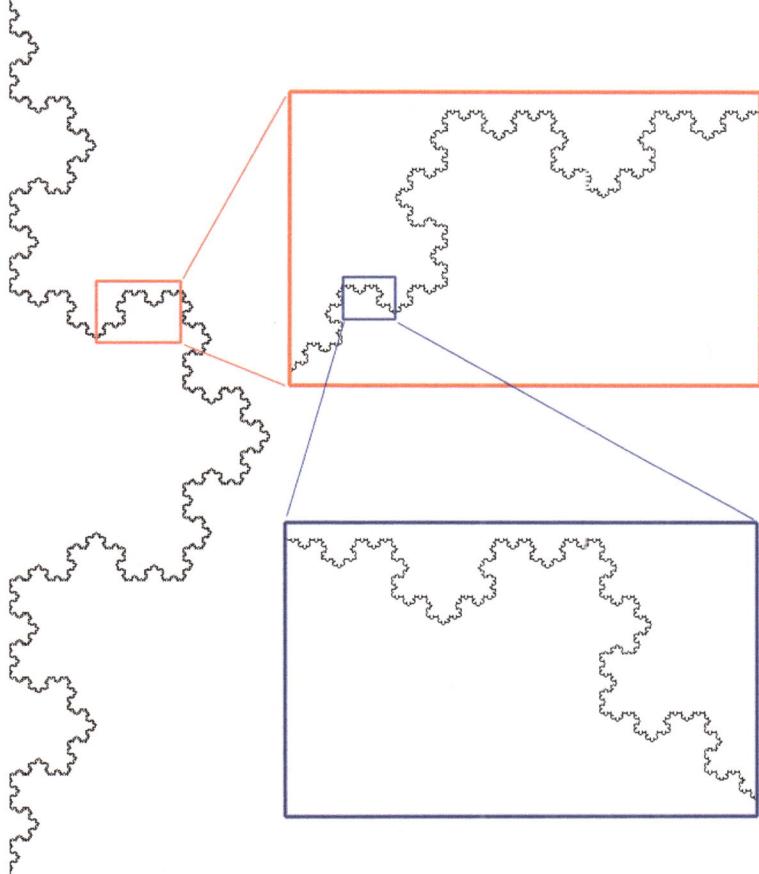

Fig. 1.1 Demonstration of geometric self-similarity

Fig. 1.2 Demonstration of
statistical self-similarity

friendly to anyone not familiar with dimensional analysis. Discussions on these topics cannot be undertaken within the scope of the present book. However, the good news is that, a formal definition of fractals is not really necessary to either learn or to appreciate their use in deciphering the latent symmetries in protein exterior organization. What however is of utmost necessity for us, is to develop a qualitative, and yet, strong conceptual ground about fractals; which will teach us how to identify the possible fractal nature in an object merely by observing its properties.

1.2 Introduction to Fractals

In his 1975 work (Mandelbrot 1975) Benoit Mandelbrot coined the term Fractal, and described it as follows:

> A [fractal is a] rough or fragmented geometric shape that can be subdivided in parts, each of which is (at least approximately) a reduced/size copy of the whole.

He derived the word from the Latin word 'fractus', that implies broken something. The word 'fractal' is a collective name for a diverse class of geometrical objects, or sets, holding most of, or all of the following properties (Falconer 1990):

1. The set will have fine structure, it will have details on arbitrary scales.
2. The set is too irregular to be described with classical Euclidean geometry, both locally and globally.
3. The set has some form of self-similarity; this self-similarity does not necessarily need to be exact (viz. geometric self-similarity), it can be approximate or statistical self-similarity also.
4. The Hausdorff dimension (Hausdorff 1919) of the set is strictly greater than its Topological dimension.
5. The set has a simple definition, i.e., it can be defined recursively.

The property (4) is Mandelbrot's original definition of a fractal, however, this property has been proven not to hold for many sets that are actually fractal. Hence, Mandelbrot's definition of fractals is no longer considered to be the most complete definition of fractals. In fact, each of the five features mentioned above have been proven not to hold for at least one fractal set. Several attempts have been made hitherto to construct a pure mathematical definition of fractals, but all have been proven unsatisfactory. We will, therefore, rather loosely, use the above properties throughout this book when talking about fractals [for further details please refer to (Falconer 1990)].

Why does Nature prefer constructing all sorts of fractal objects to run natural organizations? and that too, *so* consistently?—These are deep and difficult questions that involve exploring the interfaces of (non-equilibrium) thermodynamics, topology, chaos, and many other fields. No satisfactory, general answers to these questions have been found till now. (In fact, truth be spoken, not many scientists dare to ask these questions, let alone attempting to find an answer for them.)—All that is known is that self-similarly and self-organized structures in nature develop spontaneously from dynamical processes involved in structure formation. The displayed self-similarity can loosely be considered as a critical manifestation of the time-dependent and context-dependent 'tug-of-war' between two antagonistic properties, order and disorder. To what extent the complexity of nature is attributable to accidental randomness—is not known to us, as of now. Thus, to what extent the aforementioned 'tug-of-war' is equivalent to (confrontation-filled) coexistence of both determinism and accidental development—is not clear either. In other words, although we have learnt (quite) a bit about the schemes of description of self-similar systems and self-similar processes, we have not made much headway to understand the general reasons why nature favors them so much. [Interested readers can have a look at (Peitgen et al. 1992), for some helpful (but largely inconclusive) discussions on this line.] Anyway, remaining truthful to the prevailing scientific culture, this book will not venture into these topics either; instead, it will attempt to talk about somewhat simpler questions involving fractals. We will examine the various types of fractal properties of an object of enormous importance, the proteins. But before than that we need to take a somewhat closer look at the fractals. What they really are?

Well, let us begin at the very beginning; let us attempt to measure the length of a curve. This is a "simple" problem that we routinely come across and apparently, "solve", with ease. Typically, we take a straight ruler of length L and then starting from one end of the curve we walk the ruler along the curve for it is the entire length. In doing so, if it takes total number of N steps to traverse from one end of the curve to the other, we conclude that the total length of the curve is N-times-L. Everyday experience suggests that with a smaller (straight) ruler, it takes more number of steps to cover the length of the curve. In other words, N is proportional to 1/L. We can comprehend from our experience that for really small rulers, that is, as we go to the limit of l tending to 0, the product N·L will gradually become more and more independent of L and this product is what we term as the exact length of the curve. For example, lets measure the circumference of a circle,—a fairly straightforward problem. We take a circle with radius r and inscribe it in a regular polygon of n sides. Let the length of each side of the polygon be denoted by l. Then, the perimeter of the polygon will come out to be: $n \cdot l$. Remembering our school geometry, we note further that if the chord formed by the side of the polygon subtends an angle θ, then the number of sides of the polygon, viz. n, can be expressed as $n = \frac{2 \cdot \pi}{\theta}$ and each side l can be expressed as: $l = 2 \cdot r \cdot \sin\left(\frac{\theta}{2}\right)$. Therefore, the perimeter of the polygon comes out to be $n . l = \frac{2 \cdot \pi \cdot 2 \cdot r \cdot \sin\left(\frac{\theta}{2}\right)}{\theta} = \frac{2 \cdot \pi \cdot 2 \cdot r \cdot \sin\left(\frac{\theta}{2}\right)}{2 \cdot \left(\frac{\theta}{2}\right)}$. In the case when l becomes really small, that

is, when limit of l tends to zero, θ also tends to zero and $\sin(\theta/2)$ tends to simple $\theta/2$. The rest leaves us with known formula of perimeter of a circle.

Point is that, we could perform all of these because a circle is a "smooth curve". When mathematicians talk of a "smooth curve" they merely do not talk about a smooth looks of the curve, what they *really mean* is that if one describes a curve by $y = f(x)$, then the function $f(x)$ is differentiable everywhere. Why is differentiability, you may ask, such an important criterion?—It is important because thanks to such a condition one can express the y at any point x in the neighborhood of x_0 as $y(x) = y(x_0) + (x - x_0) \left[\frac{df}{dx} \right]_{x_0} +$ higher order terms. Then, for a sufficiently small interval of $(x - x_0)$, one ends up with $y(x) = y(x_0) + (x - x_0) \left[\frac{df}{dx} \right]_{x_0}$ which is nothing but a straight line. Thus, a curve can be approximated as a straight line at a sufficiently small interval in x, *only* if it is a "smooth curve" at that point and its neighborhood. This is the reason why, at a sufficiently small value of l, a circle could be represented with a polygon.

But what happens to the cases when such nice limits do not exist? That is, instead of being "smooth", if a curve appears rough and wiggly? Furthermore, try to envisage a situation when the "wiggly"ness of this curve persists no matter how closely you look at it. Surely, the tool-set put on display beforehand, will not find much of a use then—right? This is where we start to think about functions that are continuous at every point but differentiable nowhere, in other words, we start thinking about fractals and start to appreciate the brilliance and depth of Mandelbrot's works, something that has been talked about earlier in this chapter. Fractal geometry differs fundamentally from Euclidean ideas of geometry because it treats the Euclidean concept of "length" as a process rather than an event. Moreover, fractal geometry demands that the aforementioned "process" is controlled by a constant parameter—This idea will (hopefully) acquire a clearer shape upon completing the very next section.

1.3 Misconceptions About Fractals

Although the discussion of above introduces us to the idea of fractals, we should be guarding ourselves from some misconceptions about fractals. This job is far from being simple because misconceptions galore in the field of fractals. Here we talk about some very common misconceptions. For example,

Misconception 1: Anything that is self-similar is necessarily fractal. This is, of course, not true. A straight line is a self-similar structure. Any small part of a straight line looks exactly like a large stretch of the same straight line, which, in turn, looks like the whole straight line, and finally, all straight lines look the same—But that does not make a straight line a fractal object. Many think that the reason for straight line's not being a fractal has got to do with the fact that for a straight line, the topological dimension equals the Hausdorff dimension. While this is not wrong, such a reason appears to be a bit mechanical and does not always present the physical picture. Rather, observe the fact that for a straight line, the reduction factor can be arbitrary; while for fractal object, the reduction factor needs to be characteristic, a constant. In other words, one needs to follow a constant rule (for recursion), which ultimately produces the necessary reduction required to construct a fractal object. For example, the Koch curve, described earlier, can be reduced only by factors of $\frac{1}{3^n}$ (where n is integral), to obtain the self-similarity. For straight line, such a definite rule for recursion does not exist. Hence, do not think that all self-similar structures are fractal structures.

Misconception 2: Fractal dimension must always be fractional. It comes as surprise, but this is not true either. The Peano curve (1890) and some other fractals (the Devil's staircase (Weisstein http://mathworld.wolfram.com/DevilsStaircase.html), for example) have integer dimensions. For the Peano curve (which is a space-filling curve, that starts from a 1 dimensional straight line but applying a definite rule, gradually fills up the entire space, viz. a 2 dimensional area), the fractal dimension is 2.

Misconception 3: A bounded curve having infinite length must be a fractal. This, again, is a wrong idea. A Koch curve is bounded (that is, a Koch curve is confined within a finite region of plane) and it has infinite length. But all the curves that are bounded and that have infinite lengths may not be fractal. Take, for example, the case of a spiral. (Diagram of a simple spiral is provided in Fig. 1.3). Such a spiral is made up of circle segments of radii r_k, so that the arc length s_k of the kth circle segment comes out to be $\left(\frac{\pi}{2}\right) r_k$. Hence the length of the spiral is given by: $l = \sum_{k=1}^{\infty} s_k = \left(\frac{\pi}{2}\right) \sum_{k=1}^{\infty} r_k$ —but this expression diverges at $r_k = \frac{1}{k}$.

To clear up these misconceptions and many others, interested readers can benefit much from studying (Feder 1988; Falconer 1990; Peitgen et al. 1992).

Another important point (rather, another huge misconception) should be clarified here. The present discussion on the background of present-day fractals should not be confused with the history of fractional calculus. Details of this difference, alongside a short history of 'continuous-everywhere-but-differentiable-nowhere' are presented in my other book 'Fractal Symmetry of Protein Interior' and will not be touched upon here.

Fig. 1.3 A simple spiral

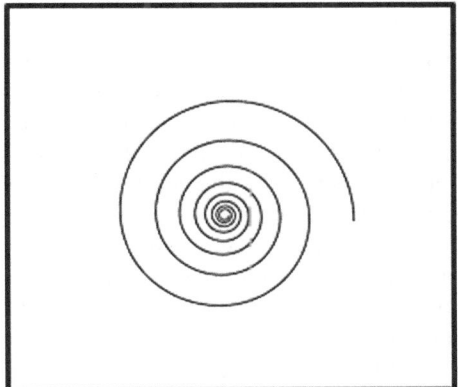

1.4 Fractal's Standing in Contemporary Science

Although the mathematical properties of fractals have been studied for over 150 years (please refer to the book 'Fractal Symmetry of Protein Interior'), fractals were a rather isolated field of mathematics until the 1970s, when, as a result of the efforts of Mandelbrot and others, many people became convinced that fractals had a number of important applications in physical and biological science that range from Brownian motion to chemical reactions—Such subdued nature of initial reactions appears a bit strange, at least from today's perspective; because the trust on symmetry in the description of natural processes and symmetry's role in deciphering the cause of natural processes—has been a guiding principle in scientific pursuits for ages. Studies that banked on symmetry under mirror reflection, rotation, translation and transformation of co-ordinates in general, are considered as the cornerstones of scientific understanding of the dynamics of natural processes. But the symmetry of self-similarity had remained hidden from the eyes of the scientists until the computers came to picture. Notably, scientists always associated symmetry with the idea of a well-ordered nature, smooth and regular. The scientific idea of beauty also revolved around these concepts of symmetries that could make objects move, grow or shrink following an orderly arrangement in space and time. In contradiction to this school of understanding of symmetries, self-similarity, that left properties of the systems invariant under the change of sizes, revealed very irregular, complex, and chaotic appearance. Predictably, therefore, it took a while before everyone could appreciate the "unusual" nature of the newfound symmetry measure. Only recently it has been established that many (but not all) of such irregular and apparently chaotic systems can be described with the help of symmetry of self-similarity.

The "subdued reaction" talked about in the last paragraph, portrayed some kind of rigidity in scientific culture. When viewed from present scenario, such rigid stances appear to be strange (if not bizarre). That is because most systems around us—trees, rivers, rocks, mountains, clouds, fire, sand-beds, coastlines, etc.—are irregular and unsmooth and they were always irregular and unsmooth. Mandelbrot's understanding of nature prompted him to think that a fractal would mathematically model a cloud better than a sphere would. The question of "How long is the Coast of Britain" may seem trivial, but when one considers the scale of our measurement—the question appears far from being simple. From a satellite orbiting around the earth in low-altitude, one can make a rough approximation; but for an ant walking the shore, following every sub-inlet, every creek and every stream as it runs deep into the British Isle, the measured length tends toward infinity. Characterizing the coastline as having a fractal dimension of 1.2, therefore, proved to be a more accurate answer to the aforementioned question. By the early 1980s, it became clear that fractal geometry is a powerful tool to describe most of the structures we encounter in our everyday life. From this point onward, various works during the past three decades have portrayed totally a different picture. Since fractal systems have a self-similar (or at times, self affine) geometry at different measurement scales, results obtained from measurements made at one scale can be used to predict geometries at other scales. This property of scale-independence has been recognized as a very useful one and is therefore used extensively in studying numerous types of natural phenomena.

While the developments in fractal geometry were taking place, similar developments were taking place in a new branch of Physics, the chaos theory. The two ideas fed each other, sharing similar roots mathematically. Thanks to the simultaneous growth and symbiosis of chaos theory and fractal geometry, the idea of a clockwork universe (a universe where a knowledge of all of the initial states of all particles and exact knowledge of the classical laws of physics would allow you to calculate the evolution of the universe from that point onwards exactly)—was brought down entirely.

Why are we studying proteins?—Well, apart from their decisive roles in performing biological functions, proteins present themselves as qualified objects to study the dynamics of complex systems. Because of the aperiodic arrangement of protein constituents, viz. the amino acids, conformational states of proteins consist of many substates with nearly the same energy. Thus potential energy surface of a protein, almost invariably, turns out to be a rough hyper-surface in a high dimensional configuration space with an immense number of local minima. Proteins share this feature with other complex systems like spin glasses, glass-forming liquids, macromolecular melts, etc. But an advantageous aspect with proteins is that, compared to other (aforementioned) systems, a protein is a relatively small (and therefore manageable) system to work with. Furthermore, proteins are identically reproduced by nature (that is, one less worry to overcome). Therefore in recent years, proteins, especially small proteins like myoglobin, have become model systems to study dynamics of complex systems.

1.5 Introduction to Proteins and Protein Exterior

Proteins are naturally occurring polypeptide chains made of some or all of 20 types of amino acids (alternatively referred to as 'residues'). Each amino acid contains a central carbon atom (C_α), which is attached to a hydrogen atom and a side-chain, constituted of an amino group (NH_2) and a carboxyl group. Thus the only difference between amino acids stems from the side-chain group attached to the C_α atom. According to the chemical nature of these side chains, the amino acids are divided into two types, hydrophobic and polar (including charged and non-charged ones).

Proteins, linear chains of amino acids arranged in a specific order, are synthesized in a cell organelle named ribosome. During protein synthesis, the amino acids are connected end-to-end into a one-dimensional sequence by the formation of peptide bonds in which the carboxyl group of one amino acid condenses with the amino group of the next and eliminates water. Protein chains thus synthesized fold into the unique tertiary three-dimensional structure characteristic of each protein—this is called the 'native structure' of a protein. This process involves a complex molecular recognition phenomenon that depends on the cooperative action of relatively weak non-bonded interactions. This book will not be talking about any of these processes; interested readers can refer to my other book 'Fractal Symmetry of Protein Interior'. The vast number of protein three-dimensional structures available in the protein data bank (PDB) (Berman et al. 2000) provides an invaluable resource to investigate structure, function, evolution, and the complex relationship among proteins.

For the purpose of elucidation of protein function, it makes more sense to investigate protein shape, protein surface shapes, and also the physico-chemical properties (such as electrostatic potential or hydrophobicity) that can be mapped on to a protein surface. While the surface physico-chemical properties have the pivotally significant role in influencing molecular interactions (Honig and Nicholls 1995), the very first step of understanding protein functionality demands construction of a correct scheme to describe surface that encapsulates protein shape.

One of the most important concepts in protein surface studies is the concept of 'accessible surface area' which in turn is dependent upon various definitions of solvent-accessible surface of a protein. In one of the commonly used schemes (Lee and Richards 1971), the outermost atoms are represented as atomic spheres having their characteristic van der Waals radii. While these concepts will be discussed later, the principal goal of this book will be to investigate the symmetry in protein surface roughness. But precisely how rough is the surface of a protein? Figure 1.4 presents us with an idea where the roughness of the binding site surface patch of a protein structure (purine nucleoside phosphorylase, structural class: α/β) is plotted, before and after binding (Banerji and Navare 2012). It is here, while attempting to quantify the roughness of surfaces, we will discover that the fractal dimension, as an index for characterization of a complex degree of natural phenomena, can also be used to describe the surface topography of a protein, accurately.

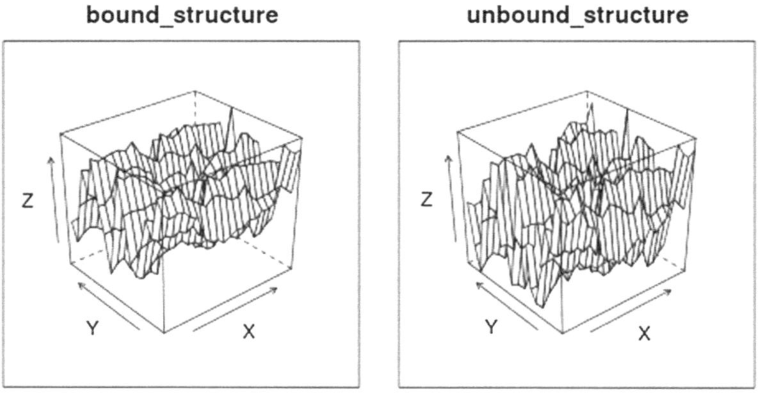

Fig. 1.4 Roughness of binding site surface patch (before and after binding). Plots of protein structures (PDB id.:1ulb and PDB id.:1ula). Upon extracting the equivalent surface patches from unbound and bound structures, plots were made by using the statistical package R

Most real surfaces have roughness at many different length scales. Proteins are soft-condensed matters and their surfaces are deformable. Thus, interaction between two proteins is basically interaction between two deformable rough surfaces. Deciphering the patterns in such interactions hold the key to understand bio-molecular recognition and biomolecular interactions. From the perspective of physics and engineering studies, they offer key to investigate contact and adhesion problems. Hence it is important to know how traditionally surface roughness is studied.

1.6 Introduction to Constructs that Investigate Surface Roughness, in General

Proteins, like many other natural substances, have a surface landscape made up of undulations that (at times) may include steep gradients. These constitute the topography of protein surface, a property that is often difficult to define with a few simple parameters but one that can have considerable impact on a protein's functionality. Due to difficulty in identifying and quantifying the (various interdependent) parameters, issues of topography assume particular importance in the case of soft condensed matters. It has been argued that for the inorganic surfaces, topography is a direct result of the nature of the material that defines it (Assender et al. 2002). Protein surface topography, being a function of evolution, is far more complex; hence, though we will start with simplistic discussions on surface roughness (most of which can as well be true on any rough natural surface), we will not persist much with these mechanistic and deterministic views. Here, right at the outset, I will like to inform the readers that not much is known about the general patterns that characterize protein surface roughness. Barring a minuscule fraction of papers that attempted to search for features in a generalized manner [like (Lewis and Rees 1985; Katchalski-Katzir et al. 1992; Pettit and Bowie 1999; Wang 2005; Baldacci et al. 2006)—for example], most papers on protein surface tend to talk about either particular cases (under particular contexts) or about protein surfaces in some dataset—often banking purely on computational prowess and not on solid theoretical foundation—This is why, in order to search for patterns in protein surface roughness we need to begin from the very beginning.

In general, a rough surface is assumed to be the result of a random process. Intuitively it may appear that an accurate estimate of the root mean square (RMS) height of the surface around some mean value of it—will be good enough to characterize the roughness of any surface. Such RMS roughness is indeed an often-used parameter. But it is inadequate on two counts. First, it fails to present any idea of the distance between the features on the surface; Second, it does not reflect the extent of anisotropy in the topography (for example, a surface with a few high-spikes may have the same RMS height as one with many low-lying features). To counter these problems one needs to consider the spatial characteristics (the RMS deviation, statistical distribution, etc.) alongside the spectral characteristics of the surfaces.

The spatial characteristics are traditionally studied with probability function $p(z)$ that denotes the probability that a point on the surface has a height equal to z. Accordingly, most of the parameters to quantify a rough surface can be observed to bear a relationship to the probability distribution function. The shape of this function presents us with information about the nature of the profile. A mathematical representation of this shape can be carried out with the use of the central moments, that is to say the moments of the function about the mean. These central moments for a distribution with zero mean are defined as:

$$\mu_v = \int_{-\infty}^{\infty} z^v p(z)\, dz$$

which, in discreet form looks like:

$$\mu_v = \frac{1}{n} \sum_{i=1}^{n} z_i^v$$

It can be seen that the first moment is zero. The second moment quantifies the variance (σ^2) of the surface heights, whereby surface height standard deviation comes out to be: $= \sqrt{\mu_2}$. Skewness of any statistical distribution measures its symmetry. For absolutely symmetrical distributions, we have skewness $= 0$. Skewness of the distribution is defined as: $\frac{\mu_3}{\mu_2^{3/2}}$. Kurtosis of any statistical distribution measures its spikiness. A normal distribution has a kurtosis equal to 3; kurtosis < 3 corresponds to a statistical distribution that is flatter than the Gaussian distribution, the opposite being true for a distribution with kurtosis > 3. Kurtosis of the distribution is defined as: $= \frac{\mu_4}{\mu_2^2}$.

But, as mentioned in the second paragraph of the present section, these constructs are far from being sufficient to quantify surface topography. To characterize the distance between the features on the surface one may resort to 2-Dimensional characterization of surface heights. While it may seem intuitive that resorting to Fourier transform may be appropriate for such 2D analysis, it has been found out that 2D autocorrelation function (ACF)—based analysis is most suitable for amorphous systems (Guinier 1963). The broad scheme for 2D ACF is given by:

$$ACF(x, y) = \frac{\iint S(x - x_1,\ y - y_1) \times S(x_1, y_1) dx_1\, dy_1}{\iint S^2(x, y) dx dy}$$

where one attempts to compare the height of a surface point (x,y) with that of another point (x_1,y_1) on the same surface by employing a texture profiling function S(x,y) on the surface (x,y). Thus, keeping the denominator as the normalization factor, the autocorrelation function ACF (x,y) compares surface heights at any two points and maps this comparison as a function of distance between them. Therefore, a surface with undulation at (more or less) regular spacing will register oscillations in ACF, a short range (local) interactions amongst correlations will register an initial decay, etc., (Goldbeck-Wood et al. 2002; Assender et al. 2002).

Though the scheme of above is helpful for obtaining a qualitative idea about surface texture profiling, for the purpose of algorithmic implementation, the discreet form is widely perceived as more useful. To describe the spectral characteristics of a surface with discreet ACF, one starts with the scheme:

$$\text{ACF}(x, y) = S\left(\lambda_x, \lambda_y\right) = E\left[z(x, y) z(x + \lambda_x, y + \lambda_y)\right]$$

where, E denotes the expectation values. Then, in discreet form, it boils down to:

$$S(p, q) = E\left[z(x, y) z(x + p, y + q)\right] = \frac{1}{(N - p)(M - q)} \sum_{k=1}^{N-p} \sum_{l=1}^{M-q} z_{k,l} z_{k+p, l+q}$$

where $S(0,0)$ is just the second moment, viz. variance (σ^2) of heights of all the surface points. While the discussion till this point seems very relevant for protein surfaces, finding relevance of the subsequent steps undertaken in ACF analyses may seem to be difficult. That is because, based on the observation that ACF of rough surfaces are mostly found to be approximately exponential for various inorganic (engineering) surfaces, the next steps of ACF involves finalizing the (arbitrarily defined) decay lengths in two directions (called 'autocorrelation lengths') that are perpendicular to each other. To what extent an analysis in these lines (involving ellipticity ratio, spectral moments, etc.) are pertinent for protein surface is debatable; but relevant or not, discussions regarding these will not be taken up in the present book. Interested readers may choose to refer to (Guinier 1963; Nayak 1971; Thomas 1982; Newland 1984; McCool 1986; Moalic et al. 1987; Majumdar and Bhushan 1990; Wu 1990; Hu and Tonder 1992; Tong and Williams 1994; Ting 1996; Assender et al. 2002; Goldbeck-Wood et al. 2002; Bakolas 2003; Ciavarella et al. 2006; Pawlus 2008; Reizer 2011) for some excellent discussions in this line. Having said that, I must emphasize that none of these references talk about protein surfaces as such; but precisely therein exists the opportunity to investigate the extent to which the knowledge acquired from Physics and Mechanical Engineering can be made relevant in the paradigm of protein surface studies.

While the aforementioned schemes are routinely used for surface roughness characterization, a closer look into them reveals that many of the aforementioned statistical parameters depend on the resolution and the scan length of the roughness-measuring instrument. In other words, many of the surface properties, measured in these ways, may not be properties of the surface alone (Majumdar and Bhushan 1990; Wu 2000). In still other words, an estimation of some of these statistical parameters may well be instrument or tool-dependent— This is a serious point, especially for investigators of protein surface, because wrong or incomplete idea about protein surface features may (at times) decide whether a drug molecule can be effective or not. Attempting to study the scale-invariant patterns of surface roughness may get around the problem of resolution-dependent instrument-driven data acquisition. It is here that we enter the relevance of fractal dimension-based (scale-invariant, self-similar) analysis of rough surfaces; it is here we enter fractal dimension-based investigation of protein surfaces and protein exterior in general,—which forms the subject matter of this book.

References

Assender H, Bliznyuk V, Porfyrakis K (2002) How surface topography relates to materials' properties. Science 297:973–976

Bachelier L (1900) Théorie de la spéculation. Annales scientifiques de l'École Normale Supérieure 17(3):21–86

Bakolas V (2003) Numerical generation of arbitrarily oriented non-Gaussian three-dimensional rough surfaces. Wear 254:546–554

Baldacci L, Golfarelli M, Lumini A, Rizzi S (2006) Clustering techniques for protein surfaces. Pattern Recogn 39:2370–2382

Banerji A, Navare C (2012) Fractal nature of protein surface roughness: a note on quantification of change of surface roughness in active sites, before and after binding, manuscript in revision, J Mol Recogn

Berman HM, Westbrook J, Feng Z, Gilliland G, Bhat TN, Weissig H, Shindyalov IN, Bourne PE (2000) The protein data bank. Nucleic Acids Res 28:235–242

Bourke P, hypertext link: http://paulbourke.net/fractals/fracdim/, last visited: 15/12/2012

Ciavarella M, Delfine V, Demelio G (2006) A "re-vitalized" Greenwood and Williamson model of elastic contact between fractal surfaces. J Mech Phys Solids 54(12):2569–2591

Falconer KJ (1990) Fractal geometry: mathematical foundations and applications. Wiley, ISBN 0-471-92287-0

Feder J (1988) Fractals. Plenum Press, New York

Goldbeck-Wood G, Bliznyuk VN, Burlakov V et al (2002) Surface structure of amorphous polystyrene: comparison of sfm imaging and lattice chain simulations. Macromolecules 35(13):5283–5289

Guinier A (1963) X-ray diffraction in crystals, imperfect crystals, and amorphous bodies. W. H. Freeman, San Francisco

Hausdorff F (1919) Dimension und ¨ausseres Mass. Math Ann 79:157–179

Honig B, Nicholls A (1995) Classical electrostatics in biology and chemistry. Science 268:1144–1149

Hu YZ, Tonder K (1992) Simulation of 3-D random rough surface by 2-D digital filter and Fourier analysis. Int J Mach Tools Manuf 32:83–90

Katchalski-Katzir E, Shariv I, Eisenstein M, Friesem AA, Aflalo C, Vakser IA (1992) Molecular surface recognition: determination of geometric fit between proteins and their ligands by correlation techniques. Proc Natl Acad Sci USA 89:2195–2199

Kaye BH (1994) A random walk through fractal dimensions. VCH, 2nd edn, ISBN 3-527-29078-8

Kolmogorov A (1941) The local structure of turbulence in incompressible viscous fluid for very large reynolds number. Comptes Rendus de l'Académie des sciences 30:9–13

Lee B, Richards FM (1971) The interpretation of protein structures: estimation of static accessibility. J Mol Biol 55:379–400

Lewis M, Rees DC (1985) Fractal surfaces of proteins. Science 230:1163–1165

Majumdar A, Bhushan B (1990) Role of fractal geometry in roughness characterization and contact mechanics of surfaces. J Tribol Trans ASME 112:205–216

Mandelbrot BB (1963) The variation of certain speculative prices. J Business 36:394–419

Mandelbrot BB (1967) How long is the coast of britain? statistical self-similarity and fractional dimension. Science 156:636–638

Mandelbrot BB (1975) Les objets fractals, forme, hasard et dimension. Flammarion, Paris

Mandelbrot BB (1982) The fractal geometry of nature. W. H. Freeman and Company, First edition. ISBN 0-7167-1186-9

McCool J (1986) Comparison of models for the contact of rough surfaces. Wear 107:37–60

Moalic H, Fitzpatrick JA, Torrance AA (1987) The correlation of the characteristics of rough surfaces with their friction coefficients. Proc Inst Mech Eng 201:321–329

Nayak PR (1971) Random process model of rough surfaces. J Lubr Technol Trans ASME 97:398–407

Newland DE (1984) An introduction to random vibration and spectral analysis, 2nd edn. Longman, London

Pawlus P (2008) Simulation of stratified surface topographies. Wear 264(5–6):457–463

Peano G (1890) Sur une courbe, qui remplit toute une aire plane. Math Ann 36(1):157–160. doi:10.1007/BF01199438

Peitgen H-O, JÄurgens H, Saupe D (1992) Fractals and chaos: new frontiers of science. Springer, New York

Pettit FK, Bowie JU (1999) Protein surface roughness and small molecular binding sites. J Mol Biol 285(4):1377–1382

Reizer R (2011) Simulation of 3D Gaussian surface topography. Wear 271(3–4):539–5433

Richardson LF (1961) The problem of contiguity: an appendix to statistic of deadly quarrels. general systems: yearbook of the society for the advancement of general systems theory. 6(139):139–187. Mich.: The Society, [1956] Society for General Systems Research, Ann Arbor

Thomas TR (1982) Rough Surfaces. Longman

Ting TCT (1996) Anisotropic Elasticity. Oxford University Press, New York 15.4

Tong WM, Williams RS (1994) Kinetics of surface growth: phenomenology, scaling, and mechanisms of smoothening and roughening. Annu Rev Phys Chem 45:401

Wang X (2005) Finding patterns on protein surfaces: algorithms and applications to protein classification. IEEE Trans Knowl Data Eng 17(8):1065–1078

Weisstein EW "Devil's Staircase". From mathworld—a wolfram web resource. http://mathworld. wolfram.com/DevilsStaircase.html

Wu J–J (1990) Spectral analysis for the effect of stylus tip curvature on measuring rough profiles. Wear 230:194–200

Wu J–J (2000) Characterization of fractal surfaces. Wear 239:36–47

Chapter 2
Characterization of Protein–Protein Interfaces, Considering Surface-Roughness and Local Shape

Abstract This chapter attempts to provide an account of works that have attempted to characterize protein–protein interaction interfaces, with fractal dimension. However, such characterization of interfaces is not solely dependent upon interface roughness. Without involving the biophysical factors, we will concentrate only on geometric characterization of these interfaces. To be specific, we will attempt to talk about a possible algorithm to quantify the changes in two parameters describing any protein–protein interaction interface; namely, the curvature of shape of protein–protein interaction interface and the surface roughness of it. One can connect these two parameters through a novel methodology, 'extended unit iterated shuffle transformation'. Results show that although the interface patch for enzyme-inhibitor interaction is flatter and smoother than the non-interfacial surface patches, absolute magnitudes of shape curvatures and surface roughness of bound interfaces are greater than what they were in unbound states of concerned entities. Trends observed on antigen–antibody interfaces are found to be somewhat contradictory to the trends observed in case of enzyme-inhibitor interfaces. Antigen–antibody interfaces, like the enzyme-inhibitor interfaces, are found to be flatter and smoother than the non-interfacial surface patches. However, unlike the enzyme-inhibitor interfaces, absolute magnitudes of shape curvatures and surface roughness of bound antigen–antibody interfaces are observed to be less than what they were in unbound states of concerned entities. Algorithm proposed in the present work could quantify the effects due to changes in two extremely important interfacial parameters, through a unified scheme.

2.1 Introduction to Protein–Protein Interaction Interfaces

Most proteins interact, at least transiently, with other protein molecules. Given that many fundamental biological processes such as antigen–antibody recognition, hormone-receptor binding, and signal transduction are regulated through association and dissociation of proteins, we need to study protein–protein interactions in details. Characterizing the physico-chemical properties of interfaces through which protein–protein interactions take place, naturally, has always been a primary aim of molecular biology. Several studies have addressed protein–protein interactions and their applications in varied paradigms, ranging from rational drug design to structure prediction of multimeric complexes. Among these, some (Katchalski-Katzir et al. 1992; Todd et al. 2002; Arkin et al. 2003; Nooren and Thornton 2003) have focused on the complementarity of chemical and structural features (shape, hydrophobic patterns, distribution of electrical charges, etc.) at the binding interface as major contributors to intersubunit interactions. These studies were useful, but they were banking on an implicit assumption that protein–protein interaction interfaces can be suitably characterized by using a single reference configuration for the interacting subunits; that is, by treating them as if they were rigid molecules. Not all studies, however, had resorted to such 'rigid molecule' framework. Due to the increasing availability of computational resources and more refined theoretical models, many other studies were made possible, which investigated the role of a complementary physical effect, namely the internal dynamics of the interacting subunits. A number of studies (Rajamani et al. 2004; Li et al. 2004; Smith et al. 2005; Yogurtcu et al. 2008) have shown that even in the absence of the partner subunits, the dynamical properties of amino acids at the known interface region can differ from those of other amino acids at the protein surface.

2.2 Why One Needs Detailed Geometric Characterization of Protein–Protein Interaction Interfaces?

As mentioned earlier, many essential cellular processes such as signal transduction, transport, cellular motion, and most regulatory mechanisms are mediated by protein–protein interaction (PPI)s. Owing to such overwhelming biological significance, PPIs have been the object of much attention; especially as they relate to interactions and associations across the entire proteome. PPIs are optimized locally (Keskin et al. 2005) and they are mediated through protein interfaces.

A huge body of knowledge about PPI interfaces has been amassed. We know that PPI sites are (mostly) hydrophobic, (mostly) planar but at times globular and protruding (Chothia and Janin 1975; Argos 1988; Jones and Thornton 1997), they are composed of relatively large surfaces with shape and electrostatic complementarity (Jones and Thornton 1996; Janin 1995, 1997)—These facts separate them clearly from enzyme catalytic sites. The catalytic site of an enzyme is a cleft, often buried, sometimes deeply. Catalytic site cleft is enzyme's workshop; it is here that the catalytic reaction occurs. In comparison to protein–protein interaction interfaces, the catalytic site house a relatively small number of amino acids that are involved in binding the substrate (and/or cofactor). Interestingly, an even smaller subset of these residues is found to be vital to the enzyme's catalytic function (Bartlett et al. 2002)—Such simple patterns in either the residue composition or in geometry are not observed for protein–protein interaction interfaces. Furthermore, it is known that these interfaces can be dynamic, facilitating binding to different proteins with diverse compositions and shapes (DeLano et al. 2000;

Ma et al. 2002; DeLano 2002; Kuhlmann et al. 2000). Indeed it has been noted that simple rules to identify protein recognition sites and prediction of energetic "hot spots" (Bogan and Thorn 1998) in protein–protein interfaces fail consistently, primarily because of the extreme diversity in shape, chemical character, and plasticity of protein–protein interfaces (Joachimiak et al. 2006). Hence, no single parameter is found to differentiate protein–protein interfaces from other surface patches with absolute confidence (Bradford and Westhead 2005).

It is observed though that, geometry is an important determinant of interfaces. Hence, shape complementarity stands out as one of the principal ingredients for all scoring functions for docking methods (Chen and Weng 2003). In varying extent, almost all the docking algorithms tend to rely on the assumption that interacting proteins have a certain degree of shape complementarity. But the exact algorithm to quantify shape complementarity varies among docking algorithms, which can either be functions based on surface curvatures or functions based on features of surface areas (Chen and Weng 2003). Probing the (elusive) connection between these, viz. functions based on interface curvatures and functions of interface surface features, therefore, assumes immense importance in the paradigm of PPI studies. An attempt to connect them for a system that is dynamic and context-dependent, however, demands an algorithm that is robust, yet sensitive.

2.3 FD-Based Characterizations of Protein–Protein Interaction Interfaces

As it was stated in the last chapter, it was known that protein surfaces are fractal in nature (Lewis and Rees 1985); furthermore, more than two decades ago it was proposed that PPI interfaces can be studied with fractal dimension (FD) as a reliable marker to quantify protein surface roughness (Aqvist and Tapia 1987). But, although promising, such a scheme (Aqvist and Tapia 1987) can be observed to have found little application in future. Possible reason behind the lack of application of this method might well be due to the absence of a suitable mathematical structure that, first, respects the fractal nature of protein surfaces and second, constructs a model to describe the change in local curvature of protein exterior (PPI interfaces in particular). Applying a recently proposed mathematical structure, 'extended unit-iterated shuffle transformation' (EU-IST) (Fujimoto and Chiba 2004), a case for this methodology can be submitted where a marker is constructed. Such a marker can connect the pattern of changes of surface roughness in enzyme-inhibitor and antigen–antibody interfaces, with the pattern of changes in the local curvature of the shapes of these two types of interface.

Neither the 'Grid'—based shape complementarity methods (Katchalski-Katzir et al. 1992), nor the 'pairwise shape complementarity' (Chen and Weng 2003) methods, take into account the fractal nature of protein surface roughness. Interestingly, the FD-based approach to describe shape changes (Bowman 1995) is radically different than those studying shape deformations in a continuous way (Sederberg and Parry 1986). While the later deforms the whole shape continuously, the former deforms the shape in a way that each subpart of the shape in all scales is deformed recursively. The EUIST-based approach (Fujimoto and Chiba 2004), owes its origin to an iterated function system studies (basically, a finite set of contraction mappings on a complete metric space) (Hutchinson 1981). With the unit-IST approach, a fractal-type repetitive structure is constructed on 'between-the-points' in the domain (atoms on interface of protein PR_1, before interacting with protein B) and those in the range (same set of atoms, after PPI). Thus, this approach ensures that local resemblance in space (viz. the scale directions) is maintained. [For further details on this, please refer to (Fujimoto and Chiba 2004)]. The EUIST-based approach elaborated in this chapter, therefore, could describe the continuous process of change of interface shape during the entire course of PPI, measuring the change in surface roughness with FD at every step.

2.4   A Small Study to Probe Antigen–Antibody and Enzyme-Inhibitor Interfaces with Surface FD

To conduct such a 'pilot–study', 10 pairs of enzyme-inhibitor complex and 10 pairs of antigen–antibody complex, as provided in the dataset of a 2003 paper (Gray et al. 2003), were taken. To obtain the most biologically relevant idea of the process, biological unit information (and not that of the asymmetric unit) for the uncomplexed and complexed units [as provided by the protein data bank (PDB) (Berman et al. 2000)], were considered. [Multifaceted significance of working with the biological units of proteins, especially while studying PPI, has been described in a recent work (Jefferson et al. 2006)].

Each of these protein complex is composed of a pair of proteins, one of which was dubbed as "receptor", while the other was called "ligand", in the paper (Gray et al. 2003) from which the dataset of the current work was taken. Ideally, to study the change in comparative surface-roughness profile between interface and non-interface regions, before and after the PPI, one should have studied the roughness of receptors and ligands both—to quantify the parameters in unbound state. However, the ligand protein molecules presented in the dataset turned out to be small sized proteins, in many (if not most) of the cases. Smallness of ligand proteins posed a problem for the calculation of surface-FD for a focused solid-angle zone (discussed later); because to calculate the surface-FD one requires having statistically significant number of atoms, which were not present in most of the cases. Furthermore, since the present work resorted to mean magnitude comparisons to assess the patterns in changes of surface roughness, a significant number of surface patches with statistically significant number of atoms were required. Ligand protein molecules, as presented in the aforementioned dataset, failed to match this constraint, consistently.

Hence the current study focused only on the surface patches of the receptor protein molecule (interface, non-interface alike), while studying the features for unbound and bound state. But in the bound state, the receptor molecule resides in the state of a complex. Hence, for the studies of 'before the PPI' states, the receptor protein molecules were considered; while, for the studies of 'after the PPI' studies, the complex protein molecules were considered. Furthermore, since a nomenclature that accommodated the receptor protein's name and PDB id., alongside that of the complex protein molecule—would have been cumbersome to make sense of, in the current study, these informations were not enlisted. The receptor protein molecule is referred to as the complex molecule in unbound state—for all the 'before the PPI' investigations. Details of the receptor and ligand protein molecules can be obtained from the reference (Gray et al. 2003).

While the implementation strategy of solid angle-based interface patch characterization is discussed later, it is important at this point to talk about the biophysical importance of resorting to such a scheme. Despite the fact that the respective geometries of PPI interfaces and catalytic sites play important roles, the geometries merely provide a helpful scaffold for the actual entities to interact, which are the amino acids present therein. Now, how to identify the residues that are participating in PPI? Though PPI interfaces are quite different than enzyme catalytic sites, in this context, it helps to recall how the catalytic site residues are characterized.

To investigate and understand the role and function of the catalytic amino acids in enzymes, the first necessary step was to define a 'catalytic residue' unambiguously. One finds such a clear-cut definition in a recent work (Holliday et al. 2007), where the definition first proposed by (Bartlett et al. 2002) was made more pointed. The (Holliday et al. 2007) scheme to define catalytic residues have two distinct parts; viz.: a catalytic residue is any residue involved in the reaction that either has direct involvement in the reaction mechanism (that is, the 'reactant residues' whose chemical structure is modified during the course of the reaction, for example, the residue is involved in covalent catalysis, electron shuttling, proton shuttling, etc.); or, has indirect but essential, involvement in the reaction mechanism [that is, the so-called spectator residues, whose chemical structure does not change during the course of the reaction—these are the residues that polarize or alter the pKa of a residue, a water molecule or part of the substrate directly involved in the reaction, affect the stereospecificity or regiospecificity of the reaction, or stabilize the reactive intermediates (either by stabilizing the transition states or the intermediates themselves, or destabilizing the ground states of the substrates)]. Thus, in order to characterize the catalytic mechanism, one requires to study both the 'reactant residues' and 'spectator residues'. Whether the residues participating in PPI can be classified in the same manner or not—needs investigation, but the idea that some of the residues can actually participate in PPI and the other adjoining ones can help the PPI without actually participating in it—seems logical. Hence, one needs a construct that present the information about the spatial location of both types of residues participating in PPI; moreover, the construct should be such that varying number of (purported) 'spectator residues' can be studied (to probe their effects on PPI), without altering the scheme significantly—The solid angle-based construct is an ideal candidate that satisfies all these requirements, hence it was resorted to.

2.4.1 Algorithms, Theoretical Basis, and Implementation Strategies

The algorithm to conduct this study comprises of two sections.

2.4.1.1 Algorithm 1st Part: Calculation Scheme for the Surface Fractals

If a given PPI is known to be involving protein PR_1 and protein PR_2, the methodology described below is presented with respect to arbitrarily chosen protein PR_1. Exactly the same methodology was implemented on protein PR_2.

Placing the center of mass of the protein at the origin of spherical polar coordinate system, solid angles of 30°, 45°, and 60° were constructed to (suitably) map the surface atoms for the entire protein. To calculate the FD of a patch of protein surface constituted by these atoms, the following scheme was implemented. (Details of implementation of this scheme can be found in the last chapter of this book.)

$$D_i = 2 - \left\{ \frac{(\log A_{VdW})_i/(\log A_{VdW})_{i-1}}{(\log R_p)_i/(\log R_p)_{i-1}} \right\} \text{ and FD} = 1/n \sum_i^n D_i$$

where $(A_{VdW})_i$ represents the Van der Waals surface area as measured by probe sphere with radius $(R_p)_i$; and FD is calculated as the mean magnitude of several sets of D_is; n being the number of such sets. Probe radius (R_p) range between 1.4 and 4.4 Å was considered, with an interval of 0.3 Å; thus n was 11. Formula stated above is the implementable form of the original (Lewis and Rees 1985) formula:

$$\text{FD} = 2 - \frac{d(\log A_{VdW})}{d(\log R_p)}$$

FD magnitude was evaluated from linear profile of A_{VdW} (ordinate) versus R_p (abscissa) graph on a log–log scale. Applying this basic scheme, contact-FD (CFD), reentrant-FD (RFD), and total-FD (TFD) were calculated for contact, reentrant and total surface description respectively, for each of the protein molecules in both unbound and complex form. In the complexed form the interface atoms reside in 'buried' state, which makes it difficult to obtain information about them directly. But by systematically studying the surface roughness of the peripheral zone of the interface patch, before and after the interaction, valuable insights about the trend of change of surface roughness could be gathered—This is exactly what has been pursued throughout the present study.

Solid angles, drawn from center-of-mass of a protein, can map the protein exterior in a unique way that helps in rigorous analyses of subtended surface patches in a focused manner. In this work, protein surfaces were mapped through a series of non-overlapping solid angles. A small solid angle, viz., a solid angle with small magnitude of both $\varphi - \theta$, can 'zoom in' on focused patches of protein surface; whereas a large solid angle (that is, solid angle with large values of $\varphi - \theta$ grid) are helpful to map large vistas of surface with minimal expenditure of time. Since the objective of the current work was to extract and quantify minuscule features of protein surfaces and interfaces, the smaller grids of solid angles were predominantly used.

However, since smaller solid angles often tend to map extremely focused areas on protein surfaces, more often than not, statistically significant number of atoms (necessary requirement for calculating the fractal dimension of the patch) could not be found in those small patches. During such cases, progressively higher grids of solid angles were employed to meet the requirement. Thus, although for every protein, a $\varphi - \theta$ grid of 15°–15° was chosen as the initial probing scheme; often, the grids 15°–30°, 30°–30°, 30°–45°, etc.,—had to be employed. In some extreme cases, the large grid, viz. 60°–60°—had to be employed also.

The grid interval for solid-angle zones for a particular protein was kept constant. That means, for any protein, if the solid angle grid $\varphi - \theta$ was observed to ensure statistically significant number of atoms for most of the surface patches, it was kept constant while mapping the surface of that protein.

2.4.1.2 Algorithm 2nd Part: Scheme to Quantify Change of Interface Shapes

Next, to identify the points on the exterior of any arbitrarily chosen protein PR_1, each point p on its shape S ($p \in S$) was assigned a unique address of the form ($a \in A_L$) (address space with L symbols). This could easily be achieved with one-to-one and onto mapping scheme $M : S \to A_L$; implying an address mapping $M(p) = a$. The surface roughness depends essentially upon positions of the atoms P ($P = \{p_i\}_{i=1}^n$ $n \geq 32$, to ensure statistical significance) that constitute the local patch of protein surface. The very same set of atoms holds the local shape of the interface for any one of the proteins (PR_1). Hence, if surface roughness is denoted by fd, it can be (trivially) noted that $fd = f(P)$. One may note that the address space defined here might not have to be merely limited to the crystallographic coordinate information; instead, it will change in its composition whenever a new property is considered (for example, the state of atomic polarizability values (Nagle 1990; Noorizadeh and Parhizgara 2005)—before, during and after the interaction might constitute three different address spaces; each of whom are dependent on atomic positional coordinates, but are not atomic positional coordinates themselves). To formalize, an attribute function φ_α was defined, such that $\varphi_\alpha : S \to Attr$ — that uniquely assigns the magnitude of relevant attribute ($\alpha \in Attr$) to a point ($p \in S$) (in other words, $\varphi_\alpha(p) = a$).

If position of every interface atom is defined in a space X, one can describe the change in the local shape of the interface of PR_1 by defining a shift vector $\vec{v}$ ($\vec{v} \in X$). A systematic analysis of $\vec{v}$ was found to be useful in monitoring every intermediate shape that the PR_1 interface may assume during PPI. This could be expressed in compact form as $\varphi_{X,\vec{v}} : S \to X$ or (from an equivalent microscopic perspective) as $\varphi_{\vec{v}}(P) \to \vec{v}$. However, since it is reasonable to view PPIs as time-dependent and context-dependent phenomena, it will be realistic to assert $\varphi_{X,\vec{v}} : S, t \to X$, or equivalently, $\varphi_{X,\vec{v}}(P, t) \to \vec{v}$; where t denoted the time duration of PPI. One notes that $\varphi_{\vec{v}}$ is continuous in S and t. (Although any fractal surface is inherently non-differentiable, evolution of the shift vector, capturing the change of local shape of PR_1 is continuous). Furthermore, evolution of $\varphi_{\vec{v}}$ is self-similar in nature. Hence, one could sum up the interface shape change process as:

$$\varphi_{X,\vec{v}}(P, t) = \varphi_{\vec{v}}(P) + \varphi_{\vec{v}}(P, t),$$

or else,

$$\varphi_{X,\vec{v}}(fd, t) = \varphi_{\vec{v}}(fd) + \varphi_{\vec{v}}(fd, t).$$

For FD calculation of the surface patch(described before), our algorithm depended (implicitly) on the solid angle-based algorithm (Connolly 1986) to calculate local shape of protein. Thus in the present scheme the range of solid angle was considered to be $(0, 2\pi)$; solid angle $(<\pi)$ denoted (subtended) surface of a locally concave shape, solid angle $(>\pi)$ denoted the (subtended) surface of a locally convex shape and solid angle $(=\pi)$ denoted a flat surface. However, during PPI, the local shape of PR_1 was found to change. While minuscule at times, the change itself was found to be a consistent trend. This implied that the cardinality of the set of atoms representing the surface patch subtended by the solid angles, might not be constant. In such a case the FD, describing the roughness of the surface patch, was found to have undergone subtle change in magnitude too. The subtle change in the magnitude of FD, in fact, could be identified as a consistent trend; because even in cases where the aforementioned cardinality was found to be invariant, FD magnitude was still found to be changing. So to describe the entire situation from an atom-centric perspective, by denoting less or more number of atoms with 'l' and 'm' respectively, the present scheme could define an EUIST for $32 \le l \le m$, such that:

1. $\mu(l, m, a) = a$, when number of atoms counted in the solid angle did not change, viz. $l = m$
2. $\mu(l, m, a) = \mu(l, m, \mu(l, m - 1, a))$, number of atoms counted in the solid angle did change.

Mean values of l and m drawn from representative subset of enzyme-inhibitor interactions or antigen–antibody interactions were considered as constant value for l and m for that subset of interactions.

With the help of basic address mapping $M(p) = a$, one could connect the top-down scheme with the bottom-up ('exact number of atom'-centric) view with a one-to-one onto mapping: $\psi(l, m, p) = M^{-1}(\mu'(l, m, M(p)))$. Thus the shift vector to describe shape transition could be expressed (more appropriately) as:

$$\varphi_{\vec{v}}(l, m, P, t) = \varphi_{\vec{v}}(P) + \varphi_{\vec{v}}\left(\psi^{-1}(l, m, P), t\right) = \varphi_{\vec{v}}(P) + \psi_{\vec{v}}^{-1}(l, m, P, t),$$
which is essentially the analogous bottom-up view of $\varphi_{X,\vec{v}}(P, t) = \varphi_{\vec{v}}(P) + \varphi_{\vec{v}}(P, t)$.

Due to sensitive nature of each of these parameters on context dependence, derivation of an analytic expression that connects P, t, fd, ν, l, m — could not be achieved. In absence of analytic evaluation of aforementioned mapping schemes, to ascertain the shift vector, the present work resorted to an empirical scheme. Here, the mean values of the pertinent parameters were calculated. With a more focused dataset (that contained, for example, enzyme-inhibitor interactions for class specific enzymes in statistically significant number), better representative mean values could have been assigned to corresponding entities and better information about shift vector transition function could be obtained.

2.4.2 Results

Results for comparative investigation between surface roughness of enzyme-inhibitor interfaces and that of the non-interfacial parts are obtained for each solid angle zone of the proteins. Similarly results for comparative investigation between surface roughness of antigen–antibody interfaces and that of the non-interfacial parts are obtained for each solid angle zone of the proteins. Data were obtained for Connolly-defined CFD and TFD. (RFD could not be calculated for many solid angles, because of the paucity of number of atoms in them. These are therefore not included in the final analyses.) While the huge bulk of (solid angle specific) raw data obtained thereby is difficult to make sense of, the mean magnitudes of surface roughness before and after the interaction (presented in a consummate way in Table 1.1) helps to decipher certain broad trends.

2.4.2.1 Comments on the General Trends

Obtained results confirmed the long held view that surface roughness magnitudes of interfaces for PPI are low. There was not a single case from the entire spectrum of enzyme-inhibitor and antigen–antibody interactions when the interface TFD was found to be greater than 2.25. This magnitude is less than what some old studies (Pfeifer et al. 1985; Dewey 1994) had suggested. Such low magnitude

Table 1.1 Summary of trends in change of interface roughness, during enzyme-inhibitor and antigen–antibody interactions

Comparative profile of surface roughness by FD	Interface regions of enzymes and inhibitors	Non-interface regions of enzymes and inhibitors	Interface regions of antigens and antibodies	Non-interface regions of antigens and antibodies
Before the interaction	2.119 ± 0.029	2.111 ± 0.033	2.114 ± 0.043	2.131 ± 0.041
After the interaction	2.150 ± 0.029	2.179 ± 0.030	2.108 ± 0.044	2.142 ± 0.041

of interface-FD could be explained [in the line of a previous study (Pfeifer et al. 1985)] as continuous compromise between two contrasting tendencies for optimization over the course of evolution. Higher FD magnitude of interfaces would have helped in maximizing the capture of substrate molecules from the bulk phase, whereas increase of speed of migration of substrate molecules along the protein surfaces would have been more probable if FD of interfaces assumed low magnitudes. Aqvist and Tapia (1987), while commenting upon the causality behind protein surface smoothness, had also emphasized on the contradiction between capture rate and diffusion. The fact that regularity of FD magnitude for protein surfaces as obtained from the present investigation is not a protein-specific feature but of statistical nature, pointed to the fact that aforementioned explanation may well be true.

The CFD magnitudes, ranging between 2.30 and 2.92, on the other hand, depicted a completely different picture. But this trend should not be considered as a contradiction to the trend observed for the TFDs. CFD magnitudes are calculated only from the contact surface roughness; therefore, they suffer from drastic changes with slightest change of probe radius (Islam and Weaver 1991). More importantly, the CFD magnitudes are merely artifacts of computational methodology. Roughness of protein surfaces owes its origin to both CFD and RFD. The RFD magnitude could not always be calculated because of the unavailability of statistically significant number of atoms in the reentrant surface of interfaces. While the CFD and TFD trends do not contradict each other, no relationship could be established among them, to predict one from another. Though a broad trend, viz., increment (or decrement) in CFD, almost always accounted for the increment (or decrement) of TFD—could be observed, some exceptions to this trend could be observed also. In other words, no proportionality could be found between CFD and TFD, at least within the scope of the current study. To minimize the scope of confusion, by 'FD' from here onwards, only the TFD will be implied.

Since FD values ranged only within 2.04–2.23, attention was frequently drawn to the magnitude of the second decimal point to identify patterns in the changing profile of FD magnitudes, both in enzyme-inhibitor and antigen–antibody studies. The words "more" or "less", "increase" and "decrease" etc., in the present report should be understood in this context. To what extent such small changes in FD values can be decisive in molecular recognition, is a debatable question. An objective quantification of the same, nevertheless, was necessary.

2.4.2.2 Surface Roughness Comparison Between Interfacial and Non-interfacial Patch Before and After PPI

This set of results was derived to identify any possible distinguishing trend in the magnitude of surface roughness that the interfacial surface patch might possess over the non-interfacial ones. As a predominant trend, FD of the interface patch for enzyme-inhibitor interactions could be observed to be more than that in non-interfacial patches in uncomplexed state. Although the difference was small, it was hardly ignorable. That is, before PPI, $FD^{enz-inh}_{interface} = 2.119 \pm 0029$

and $FD^{enz-inh}_{non-interface} = 2.111 \pm 0.033$—was observed. Studies on uncomplexed units of antigens, and antibodies, however, revealed the trend just opposite to what was observed in enzymes and inhibitors. For antigens and antibodies, the leading trend (eight out of ten cases) revealed that interfacial patches of these proteins, even in their uncomplexed forms, have smooth surfaces in comparison to non-interfacial surface patches of the same proteins. This is why, before the PPI, the magnitudes $FD^{antg-antbd}_{interface} = 2.114 \pm 0.043$ and $FD^{antg-antbd}_{non-interface} = 2.131 \pm 0.041$—could be recorded. The margin of smoothness possessed by the interfacial patches of antigens and antibodies, could be noted to be less than the margin of roughness that characterized the interfacial surface patches for enzymes and inhibitors.

Principal trend for the enzyme-inhibitor interactions after PPI suggested an increase of surface roughness of the interface (and adjoining) surface patch from its magnitude in the uncomplexed state. This could be observed in the comparison of mean magnitudes of FD for interfacial surface patch before and after PPI. While the results for surface roughness of enzyme-inhibitor interfaces before PPI has been presented above; the same for after PPI, states were recorded as: $FD^{enz-inh}_{interface} = 2.150 \pm 0.029$ and $FD^{enz-inh}_{non-interface} = 2.179 \pm 0.030$. However, for the antigen–antibody interactions, the trend was less clear-cut. Here, in four out of ten cases, FD magnitude of the interfacial surface patch could be observed to have become less in the interface (and adjoining surface patch) in the complexed form, than what it was in the uncomplexed state. For another four cases though, the interfacial FD magnitude was found to be slightly higher in the interface than what it was in the uncomplexed state. For two other cases, the interface-FD magnitudes (almost) equalled the non-interface-FD. Because of these contradictory trends, the global result of comparative roughness profile between antigen–antibody interfaces (along with their adjoining surface patches) and non-interface regions, after the PPI, viz. $FD^{antg-antbd}_{interface} = 2.108 \pm 0.044$ and $FD^{antg-antbd}_{non-interface} = 2.142 \pm 0.041$ —could not show a distinct pattern. Although the mean magnitude of FD values tends to suggest an unambiguous reduction of surface roughness of antigen and antibody interfaces, one has to admit that the trend of reduction of interface-FD is not as general as the trend of increase of interface-FD in enzyme-inhibitor interactions.

2.4.2.3 Patterns in Change of Surface Roughness (FD) in Interfacial Patches

Findings presented in the last section for before and after PPI, provide a composite picture of the process. In case of the enzyme-inhibitor interactions, the portion of the surface patch in the uncomplexed form that serves as interface during PPI, for both enzymes and inhibitors, could be observed to have (slightly) higher surface roughness than the other (non-interfacial) portions of the same molecules. Surface roughness of interfaces increased by a little margin in the course of PPI for enzyme-inhibitor interactions; however, during PPI, the roughness of the non-interfacial parts of both enzymes and inhibitors, increased by slightly bigger margin. Hence, the interfacial parts of the enzyme-inhibitor complexes appear to be flat, when observed with respect to the non-interfacial patches of the same proteins.

For the antigen–antibody interactions, such clear (and dramatic) trends could not be observed. The portion of the surface patch in the uncomplexed form that serves as interface during PPI, for both antigens and antibodies, were found to have (slightly) less surface roughness than the non-interfacial portions of the same molecules. Aforementioned roughness was observed to decrease further in the course of PPI; whereby in four out of ten cases, the interfacial roughness assumed less magnitude than that of the (non)-interfacial roughness. In another four cases interfacial roughness was found to assume higher magnitude than that of the (non)-interfacial roughness; and finally, in two cases they were found to show exactly the same magnitude, presenting ultimately an inconclusive result bereft of any tangible trend.

Since it is believed that enzymes and their inhibitors co-evolve to form interfaces with high degree of surface complementarity, reason behind the consistent trend of FD increase in interface regions of both of them, could easily be understood. Higher magnitude of surface roughness could ensure an increase in the probability of better surface complementarity for the enzyme-inhibitor interactions. However, since it is known that enzyme-inhibitor and antigen–antibody complexes represent two different classes of binding (Lawrence and Colman 1993), the same logic could not be applied on the later. The immune system produces different types of antibodies in response to an antigen, some of which bind their respective epitopes quite well while others bind poorly. Thus a particular antigen–antibody complex does not necessarily possess the best possible binding interface. Such an interpretation could help in understanding the (rather) inconclusive nature of the obtained results, when it came to antigen–antibody interactions.

Obtained results vindicated the age-old assertions that PPI interfaces are notoriously difficult to be identified keeping surface roughness as a probe. Work presented in this chapter presented the entire anatomy of the process by quantifying surface roughness of the interfacial and non-interfacial surface patches for all the involved monomers before and after the PPI. Trends observed from obtained results tend to suggest that, for all practical purposes, the magnitude of surface roughness alone, account for little help in identifying the interfacial regions. However, this difficulty could be overcome, to somewhat extent, by incorporating the shape-vector based information about trends in the change of local shape on the interfaces.

2.4.2.4 Patterns in the Magnitude of Shift Vector

From an empirical perspective, magnitude assumed by the function describing the shift vector [microscopic description: $\psi_{\vec{v}}^{-1}(l,m,P,t)\left(=\varphi_{\vec{v}}(l,m,P,t)-\varphi_{\vec{v}}(P)\right)$, macroscopic description: $\varphi_{\vec{v}}(P,t)\left(=\varphi_{X,\vec{v}}(P,t)-\varphi_{\vec{v}}(P)\right)$] was calculated from the mean magnitudes of $\varphi_{\vec{v}}(l,m,P,t)$, $\varphi_{\vec{v}}(P)$, $\varphi_{X,\vec{v}}(P,t)$ and $\varphi_{\vec{v}}(P)$ —as observed in enzyme-inhibitor and antigen–antibody cases. Positive magnitudes of shift vectors in most cases for enzyme-inhibitor interactions suggested an increment of convexity and complementary concavity in the local shape of interfacial area, associated with complex formation. The $\Delta FD_{enz–inh} = 0.031$— indicates that for uncomplexed enzymes and uncomplexed inhibitors, the curvature of the local shape that holds their interfacial surface patches, increases during interaction, albeit by a little margin. Whereas, demonstrating just the opposite trend, consistent decrease (four out of five cases) of shift vector magnitudes in case of antigen–antibody interactions symbolized a decrease of convexity and concavity that characterized the interfacial shape of the uncomplexed antigens and antibodies. As a result of such decrease (the $\Delta FD_{antgn–antbd} = -0.006$), albeit small in margin, the antigen–antibody complexes could be observed to possess flat shaped interfacial surface patches in them. The fact that shape correlations might not be an important parameter while studying antigen–antibody interactions, was pointed out long ago (Lawrence and Colman 1993; Novotný et al. 1986). Current study provided a comprehensive theoretical characterization behind such claim.

The utility of resorting to shift vector calculation became apparent with these observations; that is, even without knowing the exact (non-linear, time-dependent, context-dependent) mappings between $P, t, fd, \vec{v}, l, m$; one could easily form an idea about the nature of change in the curvature of interface from the aforementioned formulae. Although the current work considered only the final magnitudes of these parameters, one may easily implement this algorithm in molecular dynamics centric studies to obtain series of magnitudes of these parameters, which might help in obtaining an elaborate observation of the evolution of shape change during the course of an interaction. Furthermore, one notes that changes in FD magnitudes are extremely small in the current work, involving mostly the second or third decimal places. The fact that shift vectors could still reveal a change in local shape building upon these minuscule differences—suggested affirmatively that they are sensitive.

An increase or decrease of curvature (either in convexity or concavity) does not necessarily imply a change in the roughness of the surface that is holding it. However, interesting geometrical aspects of such surface patch could be detected with attempts that concentrated on particular local regions of protein exterior by mapping it with solid angles drawn from any suitably chosen reference (say, centre of mass of the protein with any suitable grid of inclination and azimuth). During the course of PPI, change in the number of atoms in the surface patch described by solid angles, becomes a non-trivial possibility. In such a case, when local curvature is changing alongside the number of atoms in surface patch subtended by the solid angle, a measure of surface roughness or a measure of local curvature alone, would have merely accounted for incomplete information. The current methodology presented an approach to connect these two. One notes the existence of another algorithm that attempts to connect local shape with the roughness of the surface that holds this shape (Banerji 2011),—however, this method fails to throw much light on the nature of change in local shape (viz. the increase or decrease of curvature of local shape etc.), which can readily be obtained from an implementation of the present one.

This work can at best be viewed as a 'pilot-project' to investigate the paradigm of protein–protein interaction interfaces from a new and rigorous theoretical and computational standpoint. While some of its findings merely vindicated the old assertions, the new computational structure proposed in this work could unite the information about surface and shape deformation study successfully. The inclusive approach of the employed methodology could ensure that protein–protein inter-action interfaces are not only investigated from studying the monomeric surface roughness, but also from the (tiny but hardly ignorable) evidences, collected from peripheral surface patches of the (buried) interface. Since the shape changes during PPI were considered and connected with the surface roughness change profile, the current methodology could present a reliable template for dissection of geometric features of PPI interfaces.

References

Aqvist J, Tapia O (1987) Surface fractality as a guide for studying protein–protein interactions. J Mol Graph 5:30–34

Argos P (1988) An investigation of protein subunit and domain interfaces. Protein Eng 2:101–113

Arkin M, Randal M, DeLano W, Hyde J, Luong T, Oslob J, Raphael D, Taylor L, Wang J, McDowell R, Wells J, Braisted A (2003) Binding of small molecules to an adaptive protein–protein interface. Proc Natl Acad Sci USA 100:1603–1608

Bartlett GJ, Porter CT, Borkakoti N, Thornton JM (2002) Analysis of catalytic residues in enzyme active sites. J Mol Biol 324:105–121

Berman HM, Westbrook J, Feng Z, Gilliland G, Bhat TN, Weissig H (2000) Shindyalov i.n., bourne p.e., the protein data bank. Nucleic Acids Res 28(1):235–242

Bogan AA, Thorn KS (1998) Anatomy of hot spots in protein interfaces. J Mol Biol 280:1–9

Bowman RL (1995) Fractal metamorphosis: a brief student tutorial. Comput Graph 19(1):157–164

Bradford JR, Westhead DR (2005) Improved prediction of protein–protein binding sites using a support vector machines approach. Bioinformatics 21(8):1487–1494

Chen R, Weng Z (2003) A novel shape complementarity scoring function for protein–protein docking, PROTEINS: structure. Funct Genet 51:397–408

Chothia C, Janin J (1975) Principles of protein–protein recognition. Nature 256:705–708

Connolly ML (1986) Measurement of protein surface shape by solid angles. J Mol Graph 4:3–6

DeLano WL, Ultsch MH, de Vos AM, Wells JA (2000) Convergent solution to binding at a protein–protein interface. Science 287:1279–1283

DeLano WL (2002) Unraveling hot spots in binding interfaces: progress and challenges. Curr Opin Struct Biol 12:14–20

Dewey TG (1994) Fractal analysis of proton exchange kinetics in lysozyme. Proc Natl Acad Sci USA 91:12101–12104

Fujimoto T, Chiba N (2004) Fractal deformation using displacement vectors and their increasing rates based on extended unit iterated shuffle transformation, "Thinking in patterns: fractals and related patterns in nature". World Scientific, Ed. Miroslav M Novak, pp 57–68

Gray JJ, Moughon S, Wang C, Schueler-Furman O, Kuhlman B, Rohl CA, Baker D (2003) Protein–protein docking with simultaneous optimization of rigid-body displacement and side-chain conformations. J Mol Biol 331:281–299

Holliday GL, Almonacid DE, Bartlett GJ, O'Boyle NM, Torrance JW, Murray-Rust P, Mitchell JBO, Thornton JM (2007) MACiE (mechanism, annotation and classification in enzymes): novel tools for searching catalytic mechanisms. Nucleic Acids Res 35:D515–D520

Hutchinson JE (1981) Fractals and self similarity. Indiana Univ Math J 30:713–747

Islam SA, Weaver DL (1991) Variation of folded polypeptide surface area with probe size. Proteins Struct Funct Genet 10(4):300–314

Janin J (1995) Principles of protein–protein recognition from structure to thermodynamics. Biochimie 77:497–505

Janin J (1997) Specific versus non-specific contacts in protein crystals. Nat Struct Biol 4:973–974

Jefferson E, Walsh T, Barton G (2006) Biological units and their effect upon the properties and prediction of protein–protein interactions. J Mol Biol 364:1118–1129

Joachimiak LA, Kortemme T, Stoddard BL, Baker D (2006) Computational design of a new hydrogen bond network and at least a 300-fold specificity switch at a protein–protein interface. J Mol Biol 361:195–208

Jones S, Thornton JM (1996) Principles of protein–protein interactions. Proc Natl Acad Sci USA 93:13–20

Jones S, Thornton JM (1997) Analysis of protein–protein interaction sites using surface patches. J Mol Biol 272:121–132

Katchalski-Katzir E, Shariv I, Eisenstein M, Friesem AA, Aflalo C, Vakser IA (1992) Molecular surface recognition: determination of geometric fit between proteins and their ligands by correlation techniques. Proc Natl Acad Sci USA 89:2195–2199

Keskin O, Mab B, Nussinov R (2005) Hot Regions in Protein–Protein Interactions: The Organization and Contribution of Structurally Conserved Hot Spot Residues. J Mol Biol 345(5):1281–1294

Kuhlmann UC, Pommer AJ, Moore GR, James R, Kleanthous C (2000) Specificity in protein–protein interactions: the structural basis for dual recognition in endonuclease colicin-immunity protein complexes. J Mol Biol 301:1163–1178

Lawrence MC, Colman PM (1993) Shape complementarity at protein–protein interfaces. J Mol Biol 234:946–950

Lewis M, Rees DC (1985) Fractal surfaces of proteins. Science 230:1163–1165

Li X, Keskin O, Ma B, Nussinov R, Liang J (2004) Protein-protein interactions: hot spots and structurally conserved residues often locate in complemented pockets that pre-organized in the unbound states: implications for docking. J Mol Biol 344:781–795

Ma B, Shatsky M, Wolfson HJ, Nussinov R (2002) Multiple diverse ligands binding at a single protein site: a matter of pre-existing populations, Protein Sci 11:184–197

Nagle JK (1990) atomic polarizability and electronegativity. J Am Chem Soc 112(12):4741–4747

Nooren IMA, Thornton JM (2003) Structural characterisation and functional significance of transient protein–protein interactions. J Mol Biol 325:991–1018

Noorizadeh S, Parhizgara M (2005) The atomic and group compressibility. J Mol Str Theochem 725(1–3):23–26

Novotný J, Handschumacher M, Haber E, Bruccoleri RE, Carlson WB, Fanning DW, Smith JA, Rose GD (1986) Antigenic determinants in proteins coincide with surface regions accessible to large probes (antibody domains). Proc Natl Acad Sci USA 83(2):226–230

Pfeifer P, Welz U, Wippermann H (1985) Fractal surface dimension of proteins: lysozyme. Chem Phys Lett 113:535–540

Rajamani D, Thiel S, Vajda S, Camacho CJ (2004) Anchor residues in protein–protein interactions. Proc Natl Acad Sci USA 101:11287–11292

Sederberg TW, Parry SR (1986) Free-form deformation of solid geometric models. ACM Comput Graph 20(4):151–160

Smith GR, Sternberg MJE, Bates PA (2005) The relationship between the flexibility of proteins and their conformational states on forming protein–protein complexes with an application to protein–protein docking. J Mol Biol 347:1077–1101

Todd A, Orengo C, Thornton J (2002) Sequence and structural differences between enzyme and nonenzyme homologs. Structure 10:1435–1451

Yogurtcu ON, Erdemli SB, Nussinov R, Turkay M, Keskin O (2008) Restricted mobility of conserved residues in protein–protein interfaces in molecular simulations. Biophys J 94:3475–3485

Chapter 3
Adhesion on Protein (and Other Rough Biomolecular) Surfaces

Abstract Surface roughness can have significant effect on the adhesion forces. The mechanical theory of adhesion talks about the effect of surface roughness on adhesion. This topic is of paramount importance to mechanobiology and to a spectrum of applicative studies where adhesions to proteins are studied. More roughness of protein surface signifies more probability of van der Waal contacts between the surface patch of protein and that of the interacting molecule. A rough binding surface may indicate potentially stronger interactions between protein and the adhesive agent. Thus, an accurate characterization of surface roughness becomes essential for applicative research of diverse types of adhesion studies. In this chapter we talk about a generalized description of fractal surfaces and their role in adhesion. Influence of surface roughness on van der Waals dispersion forces is briefly touched. But then again, this is a huge topic with illustrious history; a comprehensive account of it, therefore, is out of the scope of the present chapter. For introductory ideas, however, this chapter may be of some help. Most, if not all, of the ideas discussed in the present chapter has never been tried in the paradigm of protein surfaces, to the best of my knowledge. But this is the way ahead, because application of these ideas in proteins will be immensely beneficial for a wide range of potential applications that demand an accurate description of protein surface roughness.

3.1 What Is It That We Are Talking About?

Solid surfaces, irrespective of the method of their formation, usually present as with a common broad pattern in their morphological structure, that is, these surfaces deviate from being perfectly flat surfaces. This is especially true for the biological surfaces, because protein surfaces and surfaces of other biological macromolecules never conform to idealistic flatness. Owing to the surface roughness and owing to the short-range character of interatomic forces, contact between nominally flat surfaces remains limited to a small fraction of the apparent contact area; that is, to a multitude of discontinuous junctions randomly distributed on the surface. These contact spots, named single asperities, are responsible for all sorts of tribological properties between the two solid bodies in contact. A large body of physics research has been devoted in the past to the investigation of contact mechanics, frictional properties, and adhesion properties of rough surfaces. But almost all of these research works have been conducted in the realm of inorganic surfaces and contact mechanics on the surface of proteins is merely a field in infancy—This needs to change; because without a thorough idea of contact mechanics, friction, and adhesion on protein surfaces—investigations of mechanobiology will always be limited to heuristics-driven context-specific probing. The whole idea of the present chapter is to introduce ourselves to elementary basics of contact mechanics on inorganic surfaces, so that we can import and apply the biologically relevant subset of concepts, constructs, and tools used therein to the paradigm of protein surface contact mechanics studies. Of course, this should be the subject matter of an entire book, but we only have a chapter to talk about this topic.

Let us now amplify the last paragraph to zoom in on certain finer details in the argument. The geometric structure of rough surfaces influences a multitude of biophysical phenomena. That is because rough surfaces of biomolecules account for friction and adhesion between cellular entities (that are constantly colliding and sliding over each other). These mechanical interactions (that is, friction and adhesion) help biomolecular recognition and help in transporting energy and information. Knowledge acquired from various non-biological modeling studies on similar topics have taught us that roughness features can be defined in a wide length scale, ranging from the length of a physical sample to the molecular scale. Thus, to investigate the mechanisms of any contact phenomenon, it is necessary to characterize such multi-scale rough surface. In other words, unless and until we probe the surface structure at a length scale relevant to the phenomenon, we can never characterize it adequately and due to our ignorant/wrong characterization of the surface, we will fail to obtain a comprehensive idea about two most important processes through which biological macromolecules carry out the mechanical interactions, namely, friction and adhesion. This chapter attempts to describe protein surface roughness in such a way that it helps the readers to model mechanobiological processes.

Let us begin by talking about the ground we are trying to cover in this chapter. Moreover, without getting into details of formulae, let us start by banking only upon our "common sense". Common sense suggests that if two smooth and clean surfaces are brought to proximity, the surface molecular forces will come into play and it will then take a finite force to separate these surfaces or to cause them slide on each other. These forces are expressed in terms of surface energy per unit area, which equals the work done in separating the surfaces. Over the past five decades, the effect of surface forces on the contact configuration between solids and their roughness characteristics has been studied in great details both theoretically and experimentally. Adhesion forces refer to interacting forces between surfaces. We will start with a brief review of some of these models, keeping it in mind that these models cannot be directly transported to the realm of proteins, because proteins are not "solid objects" in the classical sense.

Before we start to talk about the effect of surface forces on the contact configuration, let us zoom in on the very first sentence of the last paragraph, where surface molecular forces between two "smooth" surfaces was talked about.

We know that such "smooth" surfaces do not exist, at least in the non-idealistic world of cytoplasmic (and in general, cellular) objects. But precisely how different are the surface forces on the contact configuration, when the "contact" is taking place between two "smooth" surfaces as opposed to the "contact"s between two real-surfaces? Taking into consideration the basic fact that most real surfaces have roughness on many different length scales, we will attempt to find answer to this and other allied questions. But before going any further, we need to introduce ourselves to mechanical characteristics of smooth and rough surfaces.

3.2 Modeling Surface Roughness (in the Context of Studying Contact Mechanics)

3.2.1 Description of the System

The process of establishing and maintaining a contact is the principal method of applying loads to a deformable body and it has been found from several engineering studies, that the resulting contact stress concentration is often the most critical point in the body. For topographically smooth surfaces the real area of contact is the same as the apparent area of contact. Real surfaces, on the contrary, invariably possess some finite degree of roughness. Since the real contacting surfaces are rough, they lead to the concentration of contact in a cluster of microscopic actual contact areas. In other words, contact between two real surfaces always occurs at or near the peaks of contacting asperities and so the real area of contact generally comes out to be much smaller than the apparent contact area. Make no mistake; even a highly polished surface has surface roughness on many different length scales. Now, bank on your common sense and think of the situation when two bodies with nominally flat surfaces are brought into contact. In such a case, *the area of real contact will typically only be a small fraction of the nominal contact area.* That is because, due to the surface roughness, when two surfaces are in contact, the contact is made at a finite number of points, where the asperities on both sides touch. (Asperity contacts, however, can have different sizes.) (Figure 3.1) Hence, it will be perfectly valid to visualize the contact regions as small areas where asperities from one solid are squeezed against asperities of the other solid. Depending on the conditions, the asperities may deform elastically or plastically. Not surprisingly therefore, during contact modeling, when the equations representing the contact of a single pair of asperities are determined, one takes into account whether the contact process refers to elastic, elastic-plastic, or completely plastic deformations. But we would not get into the details of these, interested reader may find (Pullen and Williamson 1972; Johnson 1985; Kikuchi and Oden 1985; Chang et al. 1987; Hills et al. 1993; Agrait et al. 1995; Larson et al. 1999; to name a few) to be invaluable sources of information for study in this area. For real (and therefore, rough) surfaces, however, there is an added layer of complexity. Since real surfaces have roughness, it is necessary for them to combine the effects of a large number of asperity contacts. In many (non-biological) cases it may

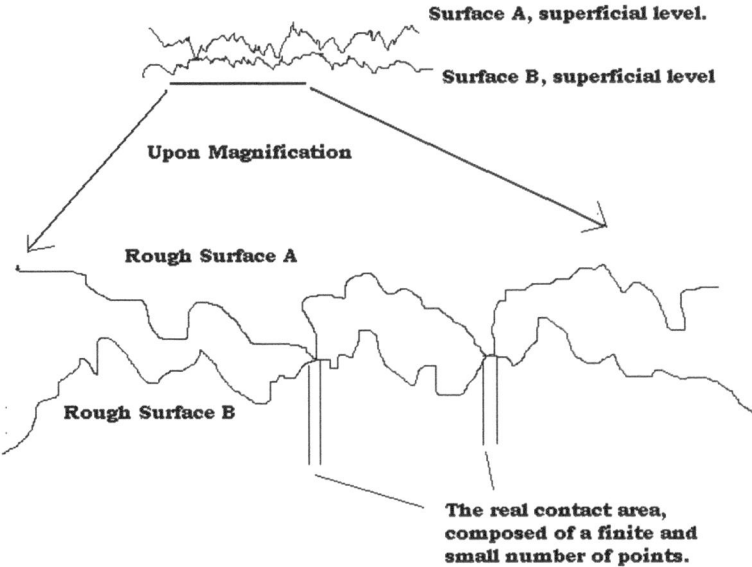

Fig. 3.1 *Schematic diagram: roughness and contact area.* Demonstrates the role that surface roughness has in limiting the contact area between surfaces. Due to the surface roughness, when two surfaces are in contact, the actual contact is made at a finite number of points, where the asperities on both sides touch. These asperity contacts have different sizes

be possible to treat these contacts as uncoupled from each other, whereas in other instances, especially in the realm of protein surfaces, the effect of coupling assumes importance. The solid-angle-based modeling of focussed surface patches as described in the last chapter, therefore, may be helpful for the present problem also.

One can, at this point attempt to define the contact problem, albeit in a loose and qualitative way. Zeroth statement of contact problem can be stated as: any point on the boundary of each interacting body in proximity, must either be in contact or not in contact. If two points are not in contact, some gap 'g' must be there between the two bodies and this gap must be positive ($g > 0$). Whereas if it is in contact, one will observe $g = 0$, by definition. A dual relationship can also be spotted here, the contact pressure, 'p', between the bodies must be positive ($p > 0$) where there is contact, and zero where there is no contact—Armed with this definition of the problem, let us now look a bit closely on the surface features.

3.2.2 A Brief Note on Surface Coupling

The usual multi-asperity contact models may be categorized as predominately uncoupled or completely coupled. Uncoupled contact models represent surface roughness as a set of asperities, often with statistically distributed parameters such as height or tip curvature. The effect of each individual asperity is assumed to be local, whereby the effect of each asperity is considered separately from other asperities. To obtain the cumulative effect one resorts to summation of the actions of individual asperities. Coupled contact problems with rough surfaces are more complicated to handle mathematically, because in these cases one needs to solve the equations of elasticity for the entire body simultaneously. This procedure leads to mixed boundary value problems, which can only be solved analytically for very simple configurations. At this point we reiterate again that one of the most tribologically important results obtained from using these asperity models is that the calculation of the true contact area, which is found to differ significantly from the nominal contact area. As stated earlier too, these quantities differ because contact between rough surfaces takes place only at and near the peaks of the asperities. It is the real contact area, which has a profound effect on biomolecular contacts.

We did mention that analytical solutions could be obtained only for a minuscule class of contact problems. Good news is that a considerable development of numerical methods can be seen in this paradigm. As a result, one finds that algorithms to handle the contact inequalities are now routinely included in many of the commercial finite element software packages.

3.2.3 Introduction to Contact Mechanics (Single Asperity and Multiple Asperity)

Heinrich Hertz, a great name of late nineteenth century physics, was probably the first to attempt the contact problem of frictionless elastic homogeneous bodies under normal loading (Hertz 1882). Despite being pioneering, Hertz's definition of system was idealistic and his assumptions were many. For example, (1) the contact area is elliptical in shape; (2) each of the contacting bodies can be approximated by an elastic half-space loaded over the plane of elliptical contact area; (3) the dimensions of the contact area must be small compared to the dimensions of each of the contacting bodies and to the radii of curvature of the surfaces; (4) the strains are sufficiently small so that constructs used for studying linear elasticity remain valid; and (5) the contact process is frictionless, so that only normal pressure is transmitted—Well, forget about the non-idealistic world of cytoplasmic objects, even for studies of (idealistic) physics, these assumptions will appear to be unrealistic. As a result, Hertz's model to describe contact process is seldom used in contemporary modeling studies.

Real surfaces are rough on the microscopic scale and the effect of roughness on the contact process, particularly in sliding contact, forms the basis of friction. Contact, as has been underlined previously too, is generally restricted to a number of microscopic 'actual contact areas' located near asperities of the rough surface. A common analytical philosophy is therefore to model the real surface as a statistical distribution of asperities of various shapes. The total load is then obtained by the sum of the individual loads on the contacting asperities, each of which is compressed a distance depending on its initial height—This basic introduction to the need behind employing statistical techniques will help us in understanding the models introduced later.

More than eighty years after Hertz's effort, the so-called GW model (Greenwood and Williamson 1966) was proposed. The Greenwood-Williamson model was used widely since its introduction. The GW model found great use in modeling the inorganic contacts. Motivated by it, various statistical (and other) models of contact were developed during the course of next 30 years from the publication of GW model. However, it is important to note that the GW model was not free of assumptions; it assumes that, in the process of contact between one rough and one smooth surface, (1) the rough surface is isotropic in nature; (2) the asperities, near to their summits, assume a spherical shape; (3) all asperity summits have the same radius of curvature, but their heights can vary randomly; and (4) there is no bulk deformation, ensuring thereby that no interaction between neighboring asperities can take place. With such a scheme of description, the total area of true contact can be expressed as:

$$A = \pi N R_{composite} \int_{d}^{\infty} (z - d)\, p(z)\mathrm{d}z$$

And the total load can be expressed as:

$$L = \frac{4}{3} E_{composite} N R_{composite}^{\frac{1}{2}} \int_{d}^{\infty} (z - d)^{\frac{3}{2}} p(z)\mathrm{d}z$$

where, A denotes the total area of true contact; L, the total load; N, the number of asperities in contact; $R_{composite}$, the composite radius of curvature; $E_{composite}$, the composite Young's modulus. Terms z and d concern probability calculation; that is, for the calculation of probability of contact at a given asperity of height z, for two surfaces that are separated by a distance d. The term $p(z)$, accordingly, denotes the probability density function of asperity heights. Two distributions of the asperity heights were considered—the exponential distribution $(p(z) = e^{-z}(when\ z > 0))$; and the Gaussian distribution $(p(z) = \frac{1}{\sqrt{2\pi}} e^{-\frac{z^2}{2}})$. It was found that the exponential distribution leads to a linear dependence of the true contact area on the applied load, whereas the Gaussian distribution yields an almost linear dependence.

Perhaps it is the considerations for identification of appropriate statistical distributions (as shown above), which resulted in a change in modeling strategy. Probably it is because of this that the models proposed after GW paper could be observed to contain strong statistical overture, where they attempted to calculate the probability of contact (say, P) at a given asperity of height z, for two surfaces that are separated by a distance d, with the (aforementioned) general scheme, viz. $P(z > d) = \int_{d}^{\infty} p(z)\,\mathrm{d}z$ By and large these models assumed some distribution laws for asperity heights and for asperity curvatures. In some cases, the other important parameters like density of surface asperities and the material and mechanical properties were also taken into account. In general, the broad consensus of results from these works suggested that if the number of asperities (N) in contact is constant and if the deformation is assumed to be elastic in nature, then the true area of contact 'A' turns out to be proportional to $L^{2/3}$, where 'L' is the applied load. In case, if the number of asperity contacts increase but the average size of each asperity contact remains the same, then A is found to be proportional to L, regardless of whether the nature of deformation is elastic or plastic. Observation of this proportionality is significant, because it allows an adhesion-based friction theory to be consistent with the observed Amontons-Coulomb friction law. We will come back to the discussion on the relationship between adhesion and friction later, when talking about cellular adhesions on extracellular matrices.

Nevertheless, adhesion properties of protein surface are of enormous practical importance. Hence, before moving over to the fractal modeling of the same, we enlist some of the prominent works that attempted to describe contact surface without using fractals, in the next section. Some of these, upon suitable modifications, may find important applications in the realm of protein adhesion (and in general, other types of biological adhesion) studies.

3.2.4 Brief Account of Contact Modeling Attempts That were Inspired by Greenwood-Williamson Model

Whitehouse and Archard (1970) attempted to model the random surface profile as a random signal characterized by suitable height distribution and an autocorrelation function. This was shown to be equivalent to asperities having a statistical distribution of both heights and radii. In the very next year, Nayak (1971) constructed an elegant statistical model that characterized a random surface by three spectral moments of the profile. These moments were asserted to be equivalent to the variances of the distribution of profile heights, slopes, and curvature, respectively. Such a scheme had lead to a distribution of peak heights that was found to be different from Gaussian. Gupta and Cook (1972) attempted to segregate the description scheme a little, whereby they permitted the tip heights to be Gaussian-distributed, but the asperity radii were modeled as log-normally distributed. Onions and Archard (1973) studied a model with a Gaussian distribution of surface heights (which is different from the asperity heights), and of asperity peak curvatures.

Four years after Nayak's work Bush et al. (1975) employed Nayak's microgeometry assumptions to develop an elastic contact model that treated asperities as elliptical paraboloids with random principal axis orientations and aspect ratio. This rigorous scheme found instant recognition; as a result, O'Callaghan and Cameron (1976) and Francis (1977) attempted to extend the Bush et al. model for the case in which both surfaces are rough and asperities are not contact at their summits. Interestingly, these investigations suggested that this type of contact is negligibly different from the original GW model.

Six years after the publication of GW model Tallian (1972) developed a model for strongly anisotropic surfaces in which the surface is modeled as a random process with Gaussian distributed heights. This study found, very significantly, that surface frequency and not just surface roughness determine the contact behavior. Meanwhile, Hisakado (1974) pointed out that a Gaussian distribution of asperity heights and curvatures for a given asperity shape may lead to a non-Gaussian distribution of the surface height. Since it is unrealistic for most engineering surfaces, to resolve this paradox, Hisakado considered a parabolic and a conical asperity shape. Here, it is worthwhile to note that non-Gaussian distribution of the surface height may truly be improbable for most inorganic surfaces, but it may find use in modeling some of the biological surfaces.

In a courageous attempt, Bush et al. (1979) attempted to model a rough surface with a random anisotropic distribution of asperity radii. Probably this trend of research was inspired by a previous attempt to investigate anisotropy. In 1976, Sayles and Thomas (1976) took a different approach and investigated a deviation from isotropy, which they referred to as "elliptical anisotropy". This term implied that contact spots have the form of randomly oriented ellipses. But Sayles-Thomas results for the contact area were found to be rather lower than that obtained with the Bush et al. model. At about the same time, in a very important piece of work, McCool (1983) investigated the limit of applicability of elastic contact models of rough surfaces, using a plane strain solution from the literature for a sinusoidally corrugated half-space. Interestingly, the range of validity of the assumptions, viz., the asperities were micro-Hertzian (that is, asperities can be approximated by a second-order polynomial in the vicinity of the contact point), and the asperities deform elastically—was shown to be related to the mean square surface slope and to the macro-contact pressure. A few years later, McCool (1986) proposed a general anisotropic model and his results demonstrated very good agreement with those obtained from simpler GW model.

The account presented here, by no means, is an elaborate history of (non-fractal) contact surface models. Interested readers can refer to (Gladwell 1980; Johnson 1985; Kikuchi and Oden 1985; Hills et al. 1993) for more details.

3.2.5 Fractal Dimension to Model Contact Surface

Finally we arrive at the fractal dimension (FD)-based description of contact modeling. Before talking about how it is done, let us talk a bit about why it is done.

Let us take a step back and put things into perspective. The GW model says that many important properties of the contact are (almost) independent of the details of the local asperity behavior, if the asperity height distribution is assumed to be Gaussian. In the special case of an exponential distribution of identical asperities, the GW model showed that the relationship between total load, thermal and electrical contact conductance and total contact area—are all linear, regardless of the constitutive law describing the contact process at the actual contact areas. (Relevance of these parameters can be understood by anyone working on mechanotransduction. We will tangentially touch by it in the end). Subsequent works, therefore, concentrated more on describing the contacting surfaces as a stationary random process.

But this classical framework was questioned in a landmark 1978 paper (Sayles and Thomas 1978) that titled 'Surface topography as a nonstationary random process'. This particular paper aside, a realization was setting in that, though useful in many engineering applications, the asperity model finds it difficult to incorporate the presence of an apparently never-ending sequence of smaller and smaller length scales. This is rather awkward because the definition of an asperity itself is scale dependent ! Thanks to this peculiar situation, for large sampling intervals one observes only a few asperities of large radius of curvature, but as the sampling interval decreases one observes more and more asperities of smaller radii. Such problems with scale-related issues (and the power law spectral density behavior) indicated strongly that a fractal description of the surface and the contact process would be more appropriate (Majumdar and Bhushan 1990; Majumdar and Bhushan 1991; Buzio et al. 2004).

As have been stated (and demonstrated) many times in this book, a number of objects or phenomena in nature exhibit inherent disorder in their very structure (organization, for phenomena). These innate disorders in them make it difficult for these objects or phenomena to surrender to a regime of idealistic and simplistic description schemes, consequently Euclidean geometry fails to characterize them. In many (but not all) such cases the structures (organization, for phenomena) of these objects and phenomena have been found to embody the fractal symmetry. Real surfaces are all rough in nature, and rough surface is one of such aforementioned phenomena, description of which is beyond the capability of idealistic constructs. That is because, if a rough surface is repeatedly magnified, increasing details of roughness keep appearing at every level of resolution and the observed profile never appears to be smooth. (Please refer to the introduction for a detailed discussion on self-similarity and smoothness). This is where we get into the realm of fractals.

Here also, like other cases mentioned in this book, we observe that ideas of fractal-like description of surface asperities were proposed before the advent of modern fractal theory. Way back in 1957, in a prophetic work, (Archard 1957) proposed a model of rough surfaces in which a progression of smaller hemispherical asperities were superposed on a larger scale, which in the limit defines what we would now describe as a fractal surface. Archard employed his model to establish that the resulting total actual contact area would be proportional to the applied load despite the nonlinearity of the Hertzian contact equations. To grasp the brilliance of the idea, we need to talk briefly about Archard's model.

Archard investigated a model of equidistant spherical asperities of the same radius R_1, which have asperities of a smaller radius R_2 on their surface, which in turn have even smaller asperities of radius R_3. Staying with the basic tenet of Hertzian elastic model, Archard calculated the dependence of the true contact zone on the load for one, two, and three sets of the asperities. The results for the contact of a plane and a sphere came out to be: $A = CP^{\frac{2}{3}}$; $A = CP^{\frac{8}{9}}$; $A = CP^{\frac{26}{27}}$ (where A denotes the true area of contact, P denotes the applied load, and C denotes a constant). For the contact of one smooth and one rough plane, the results came out to be: $A = CP^{\frac{4}{5}}; A = CP^{\frac{14}{15}}; A = CP^{\frac{44}{45}}$—These set of results suggested that these dependencies tend to converge to a linear dependence as the "order" of the asperities is increased.

Unlike statistical models, modern fractal models account for the multi-scale nature of surfaces. Fractal analysis characterizes surface roughness by two scale-independent parameters FD and G, where FD (the fractal dimension) relates to distributions of different frequencies in the surface profile and G, roughly, relates to the magnitude of variations at all frequencies.

To recapitulate a bit, a fractal surface is continuous but nondifferentiable, and it possesses self-similarity (at times, self-affinity too). These properties can be represented by the Weierstrass-Mandelbrot function, which is a superposition of sinusoids with geometrically spaced frequencies and amplitudes that follow a power law. The Weierstrass-Mandelbrot function is given by (Berry and Lewis 1980):

$$W(x) = \sum_{n=-\infty}^{\infty} \gamma^{-n(2-FD)} \left(1 - e^{i\gamma^n x}\right) e^{(i\varphi_n)}$$

where W is a complex function of the real variable x. The real portion of this equation can be used to find the fractal profile, $Z(x)$. The $Z(x)$ can describe multi-scale surface profile and is given by:

$$Z(x) = \sum_{n=-\infty}^{\infty} \gamma^{(FD-2)} \left[\cos\varphi_n - \cos\left(\gamma^n x + \varphi_n\right)\right]$$

where γ is a scaling variable that determines the density of the frequency spectrum, φ_n are phases that are randomly distributed between 0 and 2π, and FD is the fractal dimension. But we note that FD in this case must be between 1 and 2, which is not exactly what we want from a surface characterization scheme. The last equation therefore, needs to be extended to two dimensions. This is done by employing (Ausloos and Berman 1985) scheme, whereby we get:

$$Z(r,\theta) = \left(\frac{\ln\gamma}{M}\right)^{1/2} \sum_{m=1}^{M} A_m \sum_{n=-\infty}^{\infty} (k\gamma^n)^{(FD-3)}$$
$$\times \left[\cos\varphi_{m,n} - \cos(k\gamma^n r \cos(\theta - \alpha_m) + \varphi_{m,n})\right]$$

where, r and θ are standard polar coordinates. A_m represents the amplitude and can be chosen in a deterministic method or randomly. This term A_m assumes special importance. The anisotropy of the surface can be modeled suitably by adjusting the magnitude of A_m. If the surface is isotropic one will have $A_m = A = a$ constant—for all the values of m; instead, if the surface is anisotropic, A_m will vary with m. In the present case the surface is considered isotropic, whereby only one value of A_m is used. The term α_m denotes the angle corresponding to the direction of a corrugation of the surface, it can be chosen randomly. α_m can be distributed between 0 and π, or it can be distributed periodically by setting $\alpha_m = \pi m / M$, where M denotes the number of superposed ridges used to construct the surface. Parameter k is a wavenumber that can be used to scale horizontal variability in the surface. The fractal dimension, FD, now has a magnitude between 2 and 3. The term $\varphi_{m,n}$ denotes a random phase, where n is a frequency index. Finally, the term $\left(\frac{\ln \gamma}{M}\right)^{1/2}$ is a normalizing factor. Here we note that an incisive view on the fractal nature of height distributions of anisotropic rough surfaces can be found in (Blackmore and Zhou 1998)—but we will not be able to talk about this work here. If the surface is isotropic, then the fractal dimension of a surface profile, $FD_{profile}$, can be related to the fractal dimension of the surface, $FD_{surface}$, by the equation, $FD_{surface} = FD_{profile} + 1$. Furthermore, the surface and profile spectral densities can also be connected (Majumdar and Bhushan 1990). These relationships are really helpful and they allow such surfaces to be described by two parameters FD and G, as developed (quite brilliantly) by the same authors (Majumdar and Bhushan 1990). The new parameter G is the fractal roughness, which is a height scaling parameter independent of frequency. This formulation was taken further forward by Yan and Komvopoulos, when for an isotropic surface, they replaced the amplitude scale parameter A in the last equation with $A = 2\pi \left(2\pi/G\right)^{FD}$ to derive another (the more commonly known) form of the last equation (Yan and Komvopoulos 1998):

$$Z(x,y) = L \left(\frac{L}{G}\right)^{FD-2} \left(\frac{\ln \gamma}{M}\right)^{1/2} \sum_{m=1}^{M} \sum_{n=0}^{n_{max}} \gamma^{(FD-3)n}$$

$$\times \left[\cos \varphi_{m,n} - \cos\left(\frac{2\pi \gamma^n \left(x^2 + y^2\right)^{\frac{1}{2}}}{L} \times \cos\left(\tan^{-1}\left(\frac{y}{x}\right) - \frac{\pi m}{M}\right) + \varphi_{m,n}\right)\right]$$

where, the wavenumber k (discussed earlier), is expressed as $k = 2\pi/L$; L being the length of the sample. M denotes the number of superposed ridges used to construct the surface. The term $\varphi_{m,n}$ denotes a random phase, where n is a frequency index. The frequency index, n, has a lower limit of 0 and an upper limit given by: $n_{max} = \text{int} \left[\frac{\log\left(\frac{L}{L_s}\right)}{\log(\gamma)}\right]$, where L_s is the cut-off length, viz. the smallest permitted length scale, approximately 0.4 nm. Most importantly, the magnitude of parameter FD determines the contribution of high and low frequency components in the surface function Z.

Some authors tend to express the aforementioned set of equations in more compact forms. For the benefit of readers this equivalent form is given below. Here, in a straightforward manner the Weierstrass-Mandelbrot function is written as (Wang and Komvopoulos 1994):

$$Z(x) = L_x \left(\frac{G}{L_x}\right)^{FD-1} \sum_{n=0}^{M} \frac{\cos\left(2\pi \gamma^n x / L_x\right)}{\gamma^{(2-FD)n}}$$

Meaning of the symbols used in the last equation retains their meaning as explained earlier. Still, to be unambiguous, L_x denotes the fractal sample length in the x direction; G, as mentioned earlier, is the fractal roughness parameter, which is a height scaling parameter independent of frequency; FD (discussed earlier) is the fractal dimension ($1 < $ FD < 2); and γ, a scaling parameter ($\gamma > 1$), which is based on surface flatness and frequency distribution density. Based on observations of extent of surface flatness and frequency distribution it has been suggested that the magnitude $\gamma = 1.5$, models the best. The last equation simply says that the surface profile comprises a series of cosine functions with geometrically increasing frequencies from the lowest frequency $\omega_1 = \frac{1}{L_x}$, to the highest frequency $\omega_h = \frac{\gamma^M}{L_x}$. The latter is related to the smallest characteristic length L_0 (assumed to be of the order of the equilibrium atomic distance) through $\omega_h = \frac{1}{L_0}$. Hence, one finds $M = int \left[\dfrac{\log\left(L_x / L_0\right)}{\log \gamma} \right]$, where int[$\cdots$] denotes the integer part of the number in the brackets. Physically, M denotes the number of superposed ridges used to construct the surface.

The multi-scale nature of the function may now be described in terms of power spectrum of the function. The power spectrum for a self-affine fractal profile is given by (Majumdar and Tien 1990):

$$S(\omega) = \frac{G^{2(\mathrm{FD}-1)}}{2 \ln \gamma} \cdot \frac{1}{\omega^{5-2\mathrm{FD}}}$$

where, $S(\omega)$ is the power of the spectrum, (ω) being the frequency. Frequency (ω) is reciprocal to the wavelength characterizing surface roughness. The fractal parameters G and FD can be easily obtained from straight portions of a logarithmic plot of $S(\omega)$ verses (ω) (Ganti and Bhushan 1995). More importantly, the physical significance of these parameters is that as the magnitude of *FD* increases, the high frequency components become comparable in amplitude with the low frequency ones, and as magnitude of G is reduced, the roughness amplitude reduces over the entire frequency range.

We will wrap this section up with a small note on the nature of fractal description of contact areas. The Majumdar-Bhushan theory (1991) of contact for fractal surfaces was developed based on the assumption that the distribution of actual contact area sizes would be similar to that of the 'islands' generated by cutting through the surface at a constant height. Majumdar and Bhushan then obtained curvatures for the asperities so defined from random process theory and predicted the distribution of forces required to deform the asperities to the specified depth. Some years afterward, we found a more direct treatment of the fractal contact problem (Borri-Brunetto et al. 1998). This work approached the problem by first creating a finite realization of a fractal surface with the required properties and then using a numerical method to solve the resulting elastic contact problem at various levels of spatial discretization. With a coarse discretization, they obtained a few large actual contact areas, but as the grids were refined, these broke up progressively into clusters of smaller and smaller areas and the total area of actual contact decreased—These observations hold importance; they tend to suggest that in the fractal limit, the contact will consist of an infinite number of infinitesimal contact areas of total area zero. Interestingly, such a result, agrees with Archard's model, but contrasts with that of Majumdar and Bhushan, which seems to predict a fractal distribution of areas of finite size.

Tribology of fractal surfaces has grown a lot from these basic studies. Today the fractal character of surfaces grown under non-equilibrium conditions (Barabasi and Stanley 1995), fractures (Mandelbrot et al. 1984), manufactured sheet metal surfaces (Othmani and Kaminsky 1997; He and Zhu 1997), solidified liquid fronts, and in many other branches of Physics of surfaces (Pietronero and Tosatti 1996) is widely acknowledged. Other theoretical models (some of which have already been discussed here) have been developed to predict surface deformation in terms of bulk and fractal parameters (like: Majumdar and Bhushan 1991; Borodich and Mosolov 1992; Warren and Krajcinovic 1996; Yan and Komvopoulos 1998; Persson 2001; etc.). Experimentally, the contact mechanics and friction forces of fractal surfaces are studied routinely with an atomic force microscope, but we cannot get into those topics here. It is an understatement to say that tribology of protein surfaces is in infancy, but we hope that such depressing scenario will change in the coming years.

The topic of contact between roughness surfaces is vast and multifaceted. Only a small section of it could be touched here. Interested readers will benefit much by studying the following: (Hisakado 1974; McCool and Gassel 1981; Greenwood 1984; Ju and Zheng 1985; Webster and Sayles 1986; Landman et al. 1992; Powierza et al. 1992; Aramaki et al. 1993; Ren and Lee 1994; Israelachvili 2001; Tao et al. 2001).

3.2.6 Problems in Fractal Dimension-Based Modeling

Many works have discussed determining fractal parameters for use in comparing the roughness of two surfaces. But then, quantification of surface roughness and application of the roughness magnitude to various problems—talks about a wide spectrum of problems. One common complain of many researchers working on this problem with fractal constructs is that when it comes to practical implementation of the theoretical concepts, certain areas of fractal analysis are not easy to tackle. In many of the cases one needs to fall back upon 'experience', because unambiguous general implementation strategies are not always available. With the exception of the FD, and the fractal roughness G, most parameters of aforementioned equations are difficult to determine in a general practical way. The appropriate magnitudes of them, therefore, are more often than not, assumed on the basis of prior knowledge gained from working on similar problems. The suggestion that $\gamma = 1.5$ should be used, exemplify this kind of problems. Since the boundary conditions in this area of research are multifaricus, this field urgently needs a set of practical implementation strategy.

On the philosophical front one can spot another problem; the distribution of contact areas in these formulations, in general, is assumed geometrically. Such a formulation does not take account of the actual elasticity. Problem becomes more complicated when one notes that some works (Borri-Brunetto et al. 1998; Ciavarella et al. 2000; Ciavarella and Demelio 2001) have demonstrated that the actual distribution of contact areas differs by quite an extent from the one predicted by the zero-level crossing, the 'bearing area intersection' as proposed in the classical version of fractal contact theory.

3.2.7 Fractal Surfaces and Adhesion

The ability of one surface to adhere to another depends on a number of factors, such as the degree of chemical interaction between the two components, the proximity, and the area of contact. The latter two factors, quite unambiguously, depend on the topography of the surfaces to be joined. Correlation between the fractal properties of the surface and the adhesion strength has been demonstrated for polymer composites. We start our discussion on adhesion by talking about two related questions. First, how is the arrangement of atoms in surface different from that if atoms? Second, what is the thermodynamics that controls the surface interactions?—The answer to the first question introduces us to 'surface energy', which in turn introduces us to surface thermodynamic functions.

3.2.7.1 Interaction Between Surface Atoms, Lennard-Jones Formulation

Let us start with the first question. An interior atom (or interior molecule) can be defined as an atom (or molecule) within the bulk of a phase. Thus, an interior atom is always surrounded by other atoms and is attracted equiprobably in all directions. The asymmetry of the intermolecular force field on any interface emerges when the surface molecules are more strongly attracted in one direction, usually toward the bulk. As a consequence of this asymmetrical force distribution, the density of molecules in the surface differs from that in the bulk. This perturbation, more often than not, extend over many atomic spacings too.

Now, armed with this (elementary) idea about the nature of arrangement of surface atoms, let us revisit the old topic of surface contact. For this, let us look at the contacts at the really small scale, which is at the nanometer scale. In such a resolution of description of the problem, one can view the contact problem as a problem to quantify interactions between surface atomic planes, separated by some distance. In the scale of many nanometers, most of the condensed matters can still be treated as a continuum; however, in such cases, the effects of surface forces in the immediate vicinity of the contact region become important. Under such boundary conditions, the adhesive stress $\sigma(z)$ is typically represented by a special type of dispersion potential, the Lennard-Jones potential (Johnson 1998). For interactions between many inorganic substances, it has been found to assume the form:

$$\sigma\,(z) = -\frac{8(\Delta\gamma)}{3z_0}\left[\left(\frac{z}{z_0}\right)^{-3} - \left(\frac{z}{z_0}\right)^{-9}\right]$$

where z denotes the separation between atomic planes, z_0 denotes the equilibrium separation. The term $\Delta\gamma$ denotes the work of adhesion. [When two clean and smooth surfaces are brought into contact, the adhesion (or pull-off force) between them is most conveniently analyzed in terms of the work of adhesion, $\Delta\gamma$; given by: $\Delta\gamma = \gamma_1 + \gamma_2 - \gamma_{12}$, where γ_1, γ_2, and γ_{12} are the surface energies of the two interacting surfaces and the interface, respectively.]

Here it is important to note that Lennard-Jones potential between any two substances is dependent on the sub-atomic chemistry of the two substances considered. Hence, it will be wrong to expect that the same equation, as stated above, will be applicable while studying adhesion properties between proteins and other pairs of biological macromolecules. Nonetheless, though details of the expression may change from case-to-case, the general philosophical ground of the aforementioned equation will remain relevant.

Due to constraints of space, no discussion on the classic works could be undertaken here. An interested reader, however, is strongly recommended to (Bradley 1932; Johnson et al. 1971; Derjaguin et al. 1975; Tabor 1975; Tabor 1976).

Discussion of the above on the first question introduces us to the term 'surface energy', an extremely important surface thermodynamic function that forms the link between the first and the second question.

3.2.7.2 A Note on Surface Thermodynamic Functions

You may expect that talks about surface thermodynamics will start with a rigid definition of surfaces. But as it has been mentioned throughout the course of this book that, definition of surface in terms of some unique plane is very difficult (if not impossible) to obtain; especially for the target system of our study, namely, the surface of biological macromolecules (especially proteins). Thankfully, a smart way (Lewis and Randall 1961; Somorjai 1972) to get around this embarrassing situation was devised long before it was known that protein surface roughness has a fractal nature. This is done by "defining" the surface thermodynamic functions in terms of surface excess, viz. by calculating the "total" minus "bulk" magnitude of thermodynamic property concerned. For example, with this construct, the surface Gibbs potential will be defined as:

$$G^{\text{Surface}} = \left(\frac{G - G^{\text{Bulk}}}{\text{Area}}\right)$$

where 'Area' refers to the the area of the surface, G denotes the magnitude of total Gibbs potential of the system, and term G^{Bulk} refers to the magnitude that the total Gibbs potential would have assumed if all the constituent particles (atoms, molecules, etc.) were in the same state as they are in the bulk of the phase. Now, resorting to a little recollection, it was stated in Sect. 1.3.7.1. That the local environment of molecules on or near the surface is different from that of those in the bulk. This difference gives rise to an excess energy, namely, the surface energy.

Thus we notice that it to be from the bottom-up perspective (quantifying the adhesive stress with Lennard-Jones formulation) or from the top-down perspective (surface thermodynamics), quantification of adhesion process is strongly related to relevant surface energies. But surface energies are traditionally quantified in terms of energy per unit area. Therefore, one needs to be careful about what is meant by interfacial area. If the interface between phases 1 and 2 was "perfectly" flat, nobody would have had any problem in defining the interfacial area. But real-life surfaces (and protein surfaces for certain) are not flat. As a result, as the reader may have guessed unmistakably, a discussion on this question takes us back again to quantification of roughness of the interfacial area. We will talk about this topic staying within the scope of roughness of surfaces of biological macromolecules; to be more precise, from the scope of protein surface roughness. Since the aforementioned interfacial area will be comprised of a set of protein surface area, the basic scheme of quantification of extent of surface roughness by fractal dimension for interfacial area would not be different from the scheme described beforehand. However, the more important questions in this field relates to the reasons behind the observation that surface roughness affect adhesion. Stated in an even more pointed form, one may ask, why does increasing interfacial roughness often increase the extent of adhesion?—While a rigorous treatment of this question cannot be undertaken within the scope of this book, we may attempt to find a qualitative answer for it.

A classic instance of the mechanical theory of adhesion is where one phase is "keyed" into the other—This idea is intuitive and the very concept of micro-level shape-complementarity is not uncommon to the protein researchers. Indeed, it has been observed that for moderately rough surfaces, an increase in roughness (which account for increase in surface area) usually leads to a proportionate increase in adhesion, so long as the roughness does not reduce contact between the surfaces. Gent and Lai have convincingly demonstrated these effects in careful experiments with rubber adhesion (Gent and Lai 1995). But such observations, though easy to understand from common sense, does not provide us with a general theory to address the aforementioned question(s).

We note that fractal dimension-based investigations were also conducted in this paradigm. Wool (Wool 1995) had investigated the fractal nature of polymer-metal and polymer–polymer interfaces. He argued that it is the diffusion processes that often lead to fractal interfaces. It was found that though the concentration profile varies smoothly with the dimension of depth, the interface, considered in two or three dimensions, is significantly rough (Wool and Long 1993). The explanation that Polymer–polymer fractal interfaces may result from the interdiffusion of monomers or of polymers themselves, found support from the results of another set of experiments (Koizumi et al. 1994a, b). In this work, Hashimoto and his co-workers annealed the interface between polystyrene and a styrene–isoprene diblock polymer at 150 °C and showed extensive roughening of the interface by mutual interdiffusion on a micrometer scale. More useful details in this field of exploration can be found from (Eisenriegler 1993).

But diffusion-based rationalization of how increase in adhesion is caused by increase in surface roughness is not the only explanation. During the early part of 1990s another explanation could be found, which basically states that an ideally smooth surface being highly ordered would be characterized by an unfavorable extent of low entropy. In contrast, emergence and development of surface roughness can be seen as an act of increment of surface entropy in accordance with the Second Law of Thermodynamics. To gather insights into this idea, one may refer to (Koizumi et al. 1994a, b; Chen et al. 1991; Creton et al. 1994; Wool 1995).

3.2.8 Surface Roughness and Friction

Whether it is at macro-scale or at the nano-scale, the movement of an object along a surface is resisted by a certain type of force, commonly referred to as the frictional force. These forces are nonconservative and they convert the kinetic energy of the moving objects into thermal or mechanical energy. Now we know that the irreversible phenomenon of friction originates from the formation and fracture of junctions at the microscopic level formed between the rubbing surfaces, with different regimes relative to sliding velocity. But research on friction has a rich history. French physicist Guillaume Amonton (1663–1706) studied friction; he epitomized his observations in the form of an empirical law, commonly known as the law of sliding friction. This law states that for any two materials the (lateral) friction force is directly proportional to the (normal) applied load, with a constant of proportionality, the friction coefficient. That is: Friction force = friction coefficient × normal load. The parameter 'friction coefficient' is a constant and it is independent of the contact area, the surface roughness, and the sliding velocity. In most cases, the precise value of the friction coefficient depends strongly on the experimental conditions under which it is measured. Although Amonton's law is there with us for more than 300 years, no theory has yet satisfactorily explained this surprisingly general law; all attempts have been model or system dependent. A short discussion on this elementary seventeenth century law helps us to understand the inter-relationship between surface roughness, friction, and adhesion (Israelachvili 1991; Singer and Pollack 1992; Rabinowicz 1995; Persson and Tosatti 1996).

Early attempts to explain Amontons' law were based on calculations of how the microscopic surface asperities of one rough surface climb over those of the other surface, so as to allow one surface to slide past other. This straightforward formulation says that the lateral local friction force F needed to lift an asperity is equal to the normal (local) load L multiplied by $\tan\theta$, where θ denotes the maximum slope of the asperity junction. Therefore, one can readily calculate $F_i / L_i = \tan\theta_i$ at the ith asperity. Upon averaging over all asperities, one finds that the space-averaged angle θ and therefore $\tan\theta$ to be constants. In a succinct manner, one can write:

$$F = \sum F_i = \sum L_i \tan\theta_i = \langle \tan\theta_i L \rangle = \mu L$$

where, L is the total load and μ is the macroscopic friction coefficient. It is assumed that $\tan\theta_i$ and L_i are independent of each other.

The space-averaged constancy of $\tan\theta$ as the origin of friction coefficient seems easy to understand. But the last equation is not of much relevance for describing friction between biological macromolecules. That is because, where the last equation does not involve adhesion between the sliding surfaces, biological surfaces, by and large, are adhesive. To study friction on protein surfaces, therefore, we will have to resort to Derjaguin's formulation (Derjaguin 1934). This modification of Amonton's law, suitable for application on adhering surfaces is given by:

$$F = \mu (L + L_0) = \mu L + F_0$$

where, the constant internal load term, L_0, is added, so as to implicitly simulate the effect of intermolecular adhesive forces.

At this point, based on the foregoing discussion, it seems intuitive to expect that the whole situation can be expressed as: more asperities cause more roughness, and more roughness causes more friction and more adhesion. Though Derjaguin's equation can be applied on protein surfaces to study the effect of adhesive friction, no such simple tenet of logic (like, more asperity-more roughness-more friction-more adhesion) exists, at least in the non-idealistic paradigm of protein surfaces. Reason behind such assertion can be understood if a set of 10-year-old findings (Israelachvili 2001) are considered in the context of protein. The argument goes something like this: For two perfectly flat, molecularly smooth surfaces, the "real" contact area will always be the same as the "apparent" contact area. However, for rough surfaces, the real area of contact can be smaller than the apparent area (when the surfaces are hard) or can be larger than the apparent area (when the surfaces are soft). Proteins are only non-rigid condensed matters and are not classical solids, like most of the objects typically studied by surface engineers. Hence it is difficult (if not impossible) to predict in a general manner the extent to which the aforementioned 'more asperity-more roughness-more friction-more adhesion' intuitive framework will work in what regime of boundary conditions. A protein scientist working on surface friction and adhesion, therefore, needs to be cautious in his approach.

Despite the best of the efforts many topics related to contact and adhesion between two fractal surfaces could not be covered within this small space. Interested readers can refer to (Wake 1982; Ling 1989; Sahoo and Roy Chowdhury 1996; Wu 2000; Sahoo and Roy Chowdhury 2002; Packham 2003)—for excellent treatments on many of these topics.

3.3 Relevance of Studying Contact Mechanics of Biological Macromolecular Surfaces

Relevance of all of what has been discussed here is found in an emerging area of structural biology. 'Mechanobiology' is a relatively new branch of biological studies. It is an interdisciplinary study that is concerned with the cells' biological responses to *mechanical loads* and the mechanotransduction mechanisms by which these loads are transduced into a cascade of cellular and molecular events (Ingber 2006). Many studies have established that mechanical loads influence diverse array of cellular functions, such as cell proliferation, extracellular matrix (ECM) gene and protein expression, the production of soluble factors, etc. Major cellular components involved in the mechanotransduction mechanisms include the cytoskeleton, integrins, the G-proteins, receptor tyrosine kinases, mitogen-activated protein kinases, and stretch-activated ion channels. It is not possible to present an (even a vague and sketchy) overview of mechanobiology within the scope of the present book, but we can easily appreciate that the main theme of this topic revolves around the words 'mechanical loads'.

Mechanical loads induce changes in the structure, composition, and function of living tissues. We now know that mechanical forces play a fundamental role in the regulation of cell functions, including gene induction, protein synthesis, cell growth, death, and differentiation, which are essential to maintain tissue homeostasis. Conversely, abnormal mechanical loading conditions alter cellular function and change the structure and composition of the ECM, eventually leading to tissue or organ pathologies such as osteoporosis, osteoarthritis, tendinopathy, atherosclerosis, and fibrosis in the bone, cartilage, tendon, vessels, heart, lung, and skin—How do these mechanical loads actually interact with the cell (and protein) surfaces? How do we characterize the roles of different types of adhesions in recognition and transduction processes? During such mechanical load transfer operation, how does one biological macromolecule slide over the surface of another?—These questions take us to the topics discussed in the present chapter. Therein lies the relevance of discussing all those ideas that were, until now, tried and tested in the sphere of inorganic substances, the ideas that now need to be employed in the sphere of structural biology of cell surface and protein surface morphology studies.

The references: (Albelda and Buck 1990; Hynes and Lander 1992; Wang et al. 1993; Singhvi et al. 1994; Clark and Brugge 1995; Burridge and Chrzanowska-Wodnicka 1996; Shyy and Chien 1997; Lukashev and Werb 1998; Chen et al. 1997; Bissell and Nelson 1999; Folch amd Toner 2000; Cukierman et al., 2001; Hamill and Martinac, 2001; Zamir and Geiger 2001; Hamill and Martinac 2001; Seiki 2002; Sawada and Sheetz 2002; Visse and Nagase 2003; Katsumi et al. 2004; Chen et al. 2004; Hughes-Fulford 2004; Janmey and Weitz 2004)—present superb bird's eye view of evolution of mechanobiology. Though this list is preposterously undersized, readers interested about receiving the first (and yet, more-or-less complete) ideas about mechanobiology will benefit considerably by going through them.

References

Agrait N, Rubio G, Vieira S (1995) Plastic deformation in nanometer scale contacts. Langmuir 18:4505–4509

Albelda SM, Buck CA (1990) Integrins and other cell adhesion molecules. FASEB J 4(11):2868–2880

Aramaki H, Cheng HS, Chung YW (1993) The contact between roughness surfaces with longitudinal texture: I. Average contact pressure and real contact area. J. Tribology, Trans. ASME 115:419–424

Archard JF (1957) Elastic deformation and the laws of friction. Proc Royal Soc London A 243:190–205

Ausloos M, Berman DH (1985) A multivariate Weierstrass-Mandelbrot function. Proc Royal Soc London A 400:331–350

Barabasi AL, Stanley HE (1995) Fractal concepts in surface growth. Cambridge University Press, Cambridge

Berry MV, Lewis ZV (1980) On the weierstrass-mandelbrot function. Proc Roy Soc London Ser A 370:459–484

Bissell MJ, Nelson WJ (1999) Cell-to-cell contact and extracellular matrix. Integration of form and function: the central role of adhesion molecules. Curr Opin Cell Biol 11(537–539):1999

Blackmore D, Zhou JG (1998) Fractal analysis of height distributions of anisotropic rough surfaces. Fractals 6:43–58

Borodich FM, Mosolov AB (1992) Fractal roughness in contact problem. J Appl Math Mech 56:681–690

Borri-Brunetto M, Carpinteri A, Chiaia B (1998) Lacunarity of the contact domain between elastic bodies with rough boundaries. In: Frantziskonis G (ed) Probamat-21st century: probabilities and materials. Kluwer, Dordrecht, pp 45–64

Bradley RS (1932) The cohesive force between solid surfaces and the surface energy of solids. Phil Mag 13:853–862

Burridge K, Chrzanowska-Wodnicka M (1996) Focal adhesions, contractility, and signaling. Annu Rev Cell Dev Biol 12(463–518):1996

Bush AW, Gibson RD, Thomas TR (1975) The elastic contact of rough surfaces. Wear 35:87–111

Bush AW, Gibson RD, Keogh GP (1979) Strongly anisotropic rough surfaces. ASME J Lubr Tech 101:15–20

Buzio R, Malyska K, Rymuza Z, Boragno C, Biscarini F, De Mongeot FB, Valbusa U (2004) Experimental investigation of the contact mechanics of rough fractal surfaces. IEEE Trans Nanobiosci 3(1):27–31

Chang WR, Etsion I, Bogy DB (1987) An elastic–plastic model for the contact of rough surfaces. J Tribol 110:50–56

Chen CS, Mrksich M, Huang S, Whitesides GM, Ingber DE (1997) Geometric control of cell life and death. Science 276:1425–1428

Chen CS, Tan J, Tien J (2004) Mechanotransduction at cell-matrix and cell–cell contacts. Annu Rev Biomed Eng 6:275–302

Chen YL, Helm CA, Israelachvili JN (1991) Molecular mechanisms associated with adhesion and contact-angle hysteresis of monolayer surfaces. J Phys Chem 95:10736–10747

Ciavarella M, Demelio G, Barber JR, Jang YH (2000) Linear elastic contact of the Weierstrass profile. Proc R Soc London A 456:387–405

Ciavarella M, Demelio G (2001) Elastic multiscale contact of rough surfaces: archard's model revisited and comparisons with modern fractal models. J Appl Mech-Trans ASME 68:496–498

Clark EA, Brugge JS (1995) Integrins and signal transduction pathways: the road taken. Science 268:233–239

Creton C, Brown HR, Shull KR (1994) Molecular weight effects in chain pullout. Macromolecules 27:3174–3183

Cukierman E, Pankov R, Stevens DR, Yamada KM (2001) Taking cell-matrix adhesions to the third dimension. Science 294(1708–1712):2001

Derjaguin BV (1934) Molekulartheorie der €außeren Reibung. Z Phys 88:661–675

Derjaguin BV, Muller VM, Toporov YP (1975) Effect of contact deformations on the adhesion of particles. J Colloid Interface Sci 53:314–326

Eisenriegler E (1993) Polymers near surfaces: conformation properties and relation to critical phenomena. World Scientific, Singapore

Folch A, Toner M (2000) Microengineering of cellular interactions. Annu Rev Biomed Eng 2:227–256

Francis HA (1977) Application of spherical indentation mechanics to reversible and irreversible contact between rough surfaces. Wear 45:221–269

Ganti S, Bhushan G (1995) Generalized fractal analysis and its applications to engineering surfaces. Wear 180:17–34

Gent AN, Lai SM (1995) Interfacial bonding, energy dissipation and strength. Rubber Chem Technol 68:13

Gladwell GML (1980) Contact problems in the classical theory of elasticity. Sijtoff and Noordhoff, USA

Greenwood JA (1984) A unified theory of surface roughness. Proc Royal Soc London A, 393:3133–3157

Greenwood JA, Williamson JBP (1966) Contact of nominally flat surface. Proc Roy Soc London Ser A295, 300–319

Gupta PK, Cook NH (1972) Statistical analysis of mechanical interaction of rough surfaces. ASME J Lubr Tech 94:19–26

Hamill OP, Martinac B (2001) Molecular basis of mechanotransduction in living cells. Physiol Rev 81:685–740

He L, Zhu J (1997) The fractal character of processed metal surfaces. Wear 208:17–24

Hertz H (1882) Uber die beruhrung fester elastischer korper. J. Reine Angew Math 92:156–171

Hills DA, Nowell D, Sackfield A (1993) Mechanics of elastic contact. Butterworth-Heinemann Ltd, Oxford

Hisakado T (1974) Effect of surface roughness on contact between solid surfaces. Wear 28:217–234

Hughes-Fulford M (2004) Signal transduction and mechanical stress. Sci STKE (249):RE12

Hynes RO, Lander AD (1992) Contact and adhesive specificities in the associations, migrations, and targeting of cells and axons. Cell 68:303–322

Ingber DE (2006) Cellular mechanotransduction: putting all the pieces together again. FASEB J 20(7):811–827

Israelachvili JN (2001) In fundamentals of tribology bridging the gap between the macro-, micro- and nanoscales, NATO advanced science institute series. In: Bhushan B (ed) Kluwer Academic Publishers, Dordrecht, pp 631–650

Israelachvili JN (1991) Intermolecular and surface forces, 2nd edn. Academic Press, London

Janmey PA, Weitz DA (2004) Dealing with mechanics: mechanisms of force transduction in cells. Trends Biochem Sci 29(7):364–370

Johnson KL, Kendall K, Roberts AD (1971) Surface energy and the contact of elastic solids. Proc Royal Soc London A 324:301–313

Johnson KL (1985) Contact mechanics. Cambridge University Press, Cambridge

Johnson KL (1998) Mechanics of adhesion. Tribol Int 31:413–418

Ju Y, Zheng L (1985) A full numerical solution for the elastic contact of three-dimensional real rough surfaces. Johnson KLW (ed) Contact Mechanics, Cambridge University Press, Cambridge, UK

Katsumi A, Orr AW, Tzima E, Schwartz MA (2004) Integrins in mechanotransduction. J Biol Chem 279(13):12001–12004

Kikuchi N, Oden JT (1985) Contact problems in elasticity. SIAM, Philadelphia

Koizumi S, Hasegawa H, Hashimoto T (1994a) Ordered structure of block polymer/homopolymer mixtures. 5. Interplay of macro- and microphase transitions. Macromolecules 27:6532–6540

Koizumi S, Hasegawa H, Hashimoto T (1994b) Spatial distribution of homopolymers in block copolymer microdomains as observed by a combined SANS and SAXS method. Macromolecules 27:7893–7906

Landman U, Luedtke WD, Ringer EM (1992) Molecular dynamics simulations of adhesive contact formation and friction. In: Singer IL, Pollock HM (ed), Fundamentals of friction NATA ASI, Series E, vol 220, Kluwer

Larson J, Biwa S, Storakers B (1999) Inelastic flattening of rough surfaces. Mech Mat 31:29–41

Lewis GN, Randall M (1961) Thermodynamics, 2nd ed. revised by Pitzer KS, Brewer L (ed) McGraw-Hill, New York, p 472

Ling FF (1989) The possible role of fractal geometry in tribology. Tribol Trans 32:497–505

Lukashev ME, Werb Z (1998) ECM signalling: orchestrating cell behaviour and misbehaviour. Trends Cell Biol 8(11):437–441

Majumdar A, Bhushan B (1990) Role of fractal geometry in roughness characterization and contact mechanics of surfaces. J Tribol 112(2):205–216

Majumdar A, Tien CT (1990) Fractal characterization and simulation of rough surfaces. Wear 136:313–327

Majumdar A, Bhushan B (1991) Fractal model elastic–plastic contact between rough surfaces. J Tribol 113:1–11

Mandelbrot BB, Passoja DE, Paullay AJ (1984) Fractal character of fracture surfaces of metals. Nature 308:721–722

McCool JI, Gassel SS (1981) The contact of two surfaces having anisotropic roughness geometry. ASLE Spec Publ SP 7:29–38

McCool JI (1986) Comparison of models for the contact of rough surfaces. Wear 107:37–60

McCool J (1983) Limits of applicability of elastic contact models of rough surfaces. Wear 86:105–118

Nayak RP (1971) Random process model of rough surfaces. ASME J Lubrication Tech 93:398–407

Onions RA, Archard JF (1973) The contact of surfaces having a random structure. J Phys D Appl Phys 6:289–304

Othmani A, Kaminsky C (1997) Three dimensional fractal analysis of sheet metal surface. Wear 214:147–150

O'Callaghan M, Cameron MA (1976) Static contact under load between nominally flat surfaces in which deformation is purely elastic. Wear 36:79–97

Packham DE (2003) Surface energy, surface topography and adhesion. J Adhes Adhes 23(6):437–448

Physics of sliding friction (1996) Persson BNJ, Tosatti E (eds) Kluwer Academic: Dordrecht

Persson BNJ (2001) Elastoplastic contact between randomly rough surfaces. Phys Rev Lett 87(11):116101–116104

Pietronero L, Tosatti E (eds) (1996) Fractals in Physics. North Holland, Amsterdam

Powierza ZH, Klimczak T, Polijaniuk A (1992) On the experimental verification of the Greenwood–Williamson model for the contact of rough surfaces. Wear 154:115–124

Pullen J, Williamson JBP (1972) On the plastic contact of rough surfaces, Pro Roy Soc, London Ser A327, 157–173

Rabinowicz E (1995) Friction and wear of materials. Wiley, USA

Ren N, Lee SIC (1994) The effect of surface roughness and topography on the contact behavior of elastics bodies. J Tribol 116:804–811

Sahoo P, Roy Chowdhury SK (1996) A fractal analysis of adhesion at the contact between rough solids. J Eng Tribol: Proc Instn Mech Eng 210:269–279

Sahoo P, Roy Chowdhury SK (2002) A fractal analysis of adhesive wear at the contact between rough solids. Wear 253:924–934

Sawada Y, Sheetz MP (2002) Force transduction by Triton cytoskeletons. J Cell Biol 156:609–615

Sayles RS, Thomas TR (1978) Surface topography as nonstationary random process. Nature (London) 271(2):431–434

Sayles RS, Thomas TR (1976) Thermal conductance of a rough elastic contact. Appl Energy 2:249–267

Seiki M (2002) The cell surface: the stage for matrix metalloproteinase regulation of migration. Curr Opin Cell Biol 14(5):624–632

Shyy JY, Chien S (1997) Role of integrins in cellular responses to mechanical stress and adhesion. Curr Opin Cell Biol 9:707–713

Singer IL, Pollack HM (1992) Fundamentals of friction: macroscopic and microscopic processes. Kluwer, Dordrecht

Singhvi R, Kumar A, Lopez G, Stephanopoulos GN, Wang DIC, Whitesides GM (1994) Engineering cell shape and function. Science 264:696–698

Somorjai GA (1972) Principles of surface chemistry. Prentice-Hall, London

Tabor D (1975) In: Surface Physics of Materials. Blakely JM (ed) vol 2, Chapter 10. Academic Press, New York

Tabor D (1976) Surface forces and surface interactions. J Colloid Interface Sc. 58:2–13

Tallian TE (1972) The theory of partial elastohydrodynamic contact. Wear 21:49–101

Tao Q, Lee HP, Lim SP (2001) Contact mechanics of surfaces with various models of roughness descriptions. Wear 249:539–545

Visse R, Nagase H (2003) Matrix metalloproteinases and tissue inhibitors of metalloproteinases: structure, function, and biochemistry. Circ Res 92(8):827–839

Wake WC (1982) Adhesion and the formulation of adhesives, 2nd edn. Applied Science, London

Wang S, Komvopoulos K (1994) A fractal theory of the interfacial temperature distribution in the slow sliding regime. Part I. Elastic contact and heat transfer analysis. J Tribol 116:812–823

Wang N, Butler JP, Ingber DE (1993) Mechanotransduction across the cell surface and through the cytoskeleton. Science 260(5111):1124–1127

Warren TL, Krajcinovic D (1996) Random cantor set models for the elastic-perfectly plastic contact of rough surface. Wear 196:1–15

Webster MN, Sayles RS (1986) A numerical for the elastic frictionless contact of real rough surface. J Tribol 108:314–320

Whitehouse DJ, Archard JF (1970) The properties of random surfaces of significance in their contact. Proc Royal Soc London Ser A 316:97–121

Wool RP, Long JM (1993) Fractal structure of polymer interfaces. Macromolecules 26:5227–5239

Wool RP (1995) Polymer interfaces: structure and strength. Hanser, Munich

Wu JJ (2000) Characterization of fractal surfaces. Wear 239:36–47

Yan W, Komvopoulos K (1998) Contact analysis of elastic-plastic fractal surfaces. J Appl Phys 84(7):3617–3624

Zamir E, Geiger B (2001) Molecular complexity and dynamics of cell-matrix adhesions. J Cell Sci 114:3583–3590